Submerged Flow Bridge Scour
Under Clear Water Conditions

FOREWORD

This study was conducted in response to State transportation departments' requests for new design guidance to predict bridge contraction scour when the bridge is partially or fully submerged. The study included experiments at the Turner-Fairbank Highway Research Center (TFHRC) J. Sterling Jones Hydraulics Laboratory and analysis of data from Colorado State University. This report will be of interest to hydraulic engineers and bridge engineers involved in bridge foundation design. It is being distributed as an electronic document through the TFHRC Web site at http://www.fhwa.dot.gov/research/.

Jorge E. Pagán-Ortiz
Director, Office of Infrastructure
Research and Development

TECHNICAL REPORT DOCUMENTATION PAGE

1. Report No. FHWA-HRT-12-034	2. Government Accession No.	3. Recipient's Catalog No.
4. Title and Subtitle Submerged Flow Bridge Scour Under Clear Water Conditions		5. Report Date August 2012
		6. Performing Organization Code
7. Author(s) Haoyin Shan, Zhaoding Xie, Cezary Bojanowski, Oscar Suaznabar, Steven Lottes, Jerry Shen, and Kornel Kerenyi		8. Performing Organization Report No.
9. Performing Organization Name and Address GENEX SYSTEMS, LLC 2 Eaton Street, Suite 603 Hampton, VA 23669		10. Work Unit No. (TRAIS)
		11. Contract or Grant No. DTFH61-11-D-00010
12. Sponsoring Agency Name and Address Office of Infrastructure Research and Development Federal Highway Administration 6300 Georgetown Pike McLean, VA 22101-2296		13. Type of Report and Period Covered Laboratory Report May 2010–June 2012
		14. Sponsoring Agency Code

15. Supplementary Notes
The Contracting Officer's Technical Representative (COTR) was Kornel Kerenyi (HRDI-50).

16. Abstract
Prediction of pressure flow (vertical contraction) scour underneath a partially or fully submerged bridge superstructure in an extreme flood event is crucial for bridge safety. An experimentally and numerically calibrated formulation is developed for the maximum clear water scour depth in non-cohesive bed materials under different approach flow and superstructure inundation conditions. The theoretical foundation of the scour model is the conservation of mass for water combined with the quantification of the flow separation zone under the bridge deck superstructure. In addition to physical experimental data, particle image velocimetry measurements and computational fluid dynamics simulations are used to validate assumptions used in the derivation of the scour model and to calibrate parameters describing the separation zone thickness. With the calibrated model for the separation zone thickness, the effective flow depth (contracted flow depth) in the bridge opening can be obtained. The maximum scour depth is calculated by identifying the total bridge opening that creates conditions such that the average velocity in the opening, including the scour depth, is equal to the critical velocity of the bed material. Data from previous studies by Arneson and Abt and Umbrell et al. are combined with new data collected as part of this study to develop and test the proposed formulation.

17. Key Words Bridge foundations, Bridge hydraulics, Bridge scour, Submerged flow, Pressure flow, Vertical contraction	18. Distribution Statement No restrictions. This document is available to the public through the National Technical Information Service, Springfield, VA 22161.		
19. Security Classif. (of this report) Unclassified	20. Security Classif. (of this page) Unclassified	21. No. of Pages 51	22. Price

Form DOT F 1700.7 (8-72) Reproduction of completed page authorized

SI* (MODERN METRIC) CONVERSION FACTORS

APPROXIMATE CONVERSIONS TO SI UNITS

Symbol	When You Know	Multiply By	To Find	Symbol
LENGTH				
in	inches	25.4	millimeters	mm
ft	feet	0.305	meters	m
yd	yards	0.914	meters	m
mi	miles	1.61	kilometers	km
AREA				
in^2	square inches	645.2	square millimeters	mm^2
ft^2	square feet	0.093	square meters	m^2
yd^2	square yard	0.836	square meters	m^2
ac	acres	0.405	hectares	ha
mi^2	square miles	2.59	square kilometers	km^2
VOLUME				
fl oz	fluid ounces	29.57	milliliters	mL
gal	gallons	3.785	liters	L
ft^3	cubic feet	0.028	cubic meters	m^3
yd^3	cubic yards	0.765	cubic meters	m^3
NOTE: volumes greater than 1000 L shall be shown in m^3				
MASS				
oz	ounces	28.35	grams	g
lb	pounds	0.454	kilograms	kg
T	short tons (2000 lb)	0.907	megagrams (or "metric ton")	Mg (or "t")
TEMPERATURE (exact degrees)				
°F	Fahrenheit	5 (F-32)/9 or (F-32)/1.8	Celsius	°C
ILLUMINATION				
fc	foot-candles	10.76	lux	lx
fl	foot-Lamberts	3.426	candela/m^2	cd/m^2
FORCE and PRESSURE or STRESS				
lbf	poundforce	4.45	newtons	N
lbf/in^2	poundforce per square inch	6.89	kilopascals	kPa

APPROXIMATE CONVERSIONS FROM SI UNITS

Symbol	When You Know	Multiply By	To Find	Symbol
LENGTH				
mm	millimeters	0.039	inches	in
m	meters	3.28	feet	ft
m	meters	1.09	yards	yd
km	kilometers	0.621	miles	mi
AREA				
mm^2	square millimeters	0.0016	square inches	in^2
m^2	square meters	10.764	square feet	ft^2
m^2	square meters	1.195	square yards	yd^2
ha	hectares	2.47	acres	ac
km^2	square kilometers	0.386	square miles	mi^2
VOLUME				
mL	milliliters	0.034	fluid ounces	fl oz
L	liters	0.264	gallons	gal
m^3	cubic meters	35.314	cubic feet	ft^3
m^3	cubic meters	1.307	cubic yards	yd^3
MASS				
g	grams	0.035	ounces	oz
kg	kilograms	2.202	pounds	lb
Mg (or "t")	megagrams (or "metric ton")	1.103	short tons (2000 lb)	T
TEMPERATURE (exact degrees)				
°C	Celsius	1.8C+32	Fahrenheit	°F
ILLUMINATION				
lx	lux	0.0929	foot-candles	fc
cd/m^2	candela/m^2	0.2919	foot-Lamberts	fl
FORCE and PRESSURE or STRESS				
N	newtons	0.225	poundforce	lbf
kPa	kilopascals	0.145	poundforce per square inch	lbf/in^2

*SI is the symbol for the International System of Units. Appropriate rounding should be made to comply with Section 4 of ASTM E380.
(Revised March 2003)

TABLE OF CONTENTS

LIST OF FIGURES

LIST OF TABLES

LIST OF ABBREVIATIONS AND SYMBOLS

Abbreviations

3D	Three-dimensional
CFD	Computational fluid dynamics
FHWA	Federal Highway Administration
MicroADV	Micro-acoustic doppler velocimeter
PIV	Particle image velocimetry
RMSE	Root mean square error
TFHRC	Turner-Fairbank Highway Research Center

Symbols

α, β	Optimization parameters
D_{50}	Median grain size, ft
g	Gravitational acceleration, ft/s^2
h	Flow depth, ft
h_b	Vertical bridge opening height before scour, ft
h_c	Flow contraction depth at the point of maximum scour, ft
h_u	Approach flow depth, ft
h_{ue}	Effective approach flow depth directed under the bridge, ft
h_t	Flow depth above the bottom of the bridge superstructure, ft
h_w	Depth of weir flow overtopping bridge, ft
K	Variable related to bridge superstructure geometry, dimensionless
q	Unit discharge, ft^2/s
q_B	Unit discharge blocked by bridge superstructure, ft^2/s
s	Specific gravity of sediment, dimensionless
t	Separation zone thickness, ft
T	Deck thickness, ft
V	Average flow velocity, ft/s
V_b	Average velocity under the bridge before scour, ft/s
V_c	Critical velocity for the bed material, ft/s
V_u	Average approach (upstream) velocity, ft/s
V_{ue}	Effective average approach (upstream) velocity directed under the bridge, ft/s

y_2 Effective depth at point of maximum scour, ft

y_s Equilibrium scour depth, ft

ρ Density of water, slug/ft^3

μ Viscosity of water, lb-s/ft^2

CHAPTER 1. INTRODUCTION

Pressure flow (also known as vertical contraction) scour occurs when a bridge deck is insufficiently high such that the bridge superstructure becomes a barrier to the flow, causing the flow to vertically contract as it passes under the deck. A bridge deck is considered partially submerged when the lowest structural element of the bridge is in contact with the flowing water but the water is not sufficiently high to overtop the bridge deck. It is considered fully submerged when a portion of the flow overtops the bridge deck.

Pressure flow generally only occurs in extreme flood events, but these types of events are relevant for estimation of scour. When flow is sufficiently high so that it begins to approach the elevation of the bridge deck, some of the flow may be diverted laterally to the bridge approaches. Since the bridge approaches are often lower than the bridge deck, this diversion may reduce the scour potential under the bridge. Designers must evaluate the effects of scour under the bridge as well as potential damage caused by flow diversion.

An experimentally and numerically calibrated scour model was developed in this study to calculate the maximum clear water scour depth in non-cohesive bed materials under different deck inundation conditions. The theoretical formulation of the model is based on the conservation of mass of the water passing underneath the bridge deck. Particle image velocimetry (PIV) measurements and computational fluid dynamics (CFD) simulations were used to validate assumptions used in the derivation and verify calibration of parameters included in the scour model. As one of the important parameters in the pressure flow scour model, the separation zone thickness in the bridge opening was formulated analytically, calibrated experimentally, and verified by PIV and CFD analyses. The maximum scour depth was calculated by identifying the total bridge opening that resulted in the average velocity in the opening that is equal to the critical velocity of the bed material.

Experimental data were developed for this study at the Federal Highway Administration's (FHWA) Turner-Fairbank Highway Research Center (TFHRC). In addition, data from previous studies by Arneson and Abt as well as Umbrell et al. were retrieved to support the evaluation of the proposed pressure flow scour model.[1,2]

This report summarizes a literature review on pressure scour and describes the physical and theoretical foundation for the model formulation. The newly collected flume data as well as PIV and CFD analyses are summarized along with the data collected by Arneson and Abt as well as Umbrell et al.[1,2] The model formulation is refined, tested, and compared to other approaches used to estimate pressure scour. Recommendations for model application are also provided.

CHAPTER 2. LITERATURE REVIEW

Investigations on submerged flow bridge scour have been reported by Arneson and Abt, Umbrell et al., and Lyn.[1–3] Arneson and Abt conducted a series of flume tests at Colorado State University and proposed the regression equation in figure 1.[1]

$$\frac{y_s}{h_u} = -5.08 + 1.27\left(\frac{h_u}{h_b}\right) + 4.44\left(\frac{h_b}{h_u}\right) + 0.19\left(\frac{V_b}{V_c}\right)$$

Figure 1. Equation. Arneson and Abt equation for maximum equilibrium scour.[1]

Where:

y_s = Maximum equilibrium scour depth, ft.
h_u = Approach flow depth, ft.
h_b = Vertical bridge opening height before scour, ft.
V_b = Average velocity of the flow through the bridge opening before scour occurs, ft/s.
V_c = Critical velocity of the bed material in the bridge opening, ft/s.

V_c is defined by Arneson and Abt as shown in figure 2 as follows:[1]

$$V_c = 1.52\sqrt{g(s-1)D_{50}}\left(\frac{h_u}{D_{50}}\right)^{1/6}$$

Figure 2. Equation. Critical velocity per Arneson and Abt.[1]

Where:

g = Gravitational acceleration, ft/s^2.
s = Specific gravity of sediment, dimensionless.
D_{50} = Median diameter of the bed materials, ft.

Umbrell et al. conducted a series of flume tests in the FHWA TFHRC J. Sterling Jones Hydraulics Laboratory.[2] Using the mass conservation law and assuming that the velocity under a bridge at scour equilibrium is equal to the critical velocity of the upstream flow, they presented the equation in figure 3.

$$\frac{y_s + h_b}{h_u} = \frac{V_u}{V_c}\left(1 - \frac{h_w}{h_u}\right)$$

Figure 3. Equation. Umbrell et al. equation for maximum equilibrium scour.[2]

Where:

V_u = Average approach flow velocity, ft/s.
h_w = Depth of weir flow overtopping bridge, ft.

V_u must be less than or equal to V_c to insure the clear water scour assumption. When flow does not overtop the bridge deck (i.e., partially submerged flow), then h_w equals zero.

Umbrell et al. modified the previous equation to improve the fit to their laboratory data, resulting in the equation in figure 4. As part of the refinement in this equation, V_c is estimated by the equation in figure 2, but the coefficient 1.52 is replaced by 1.58.

$$\frac{y_s + h_b}{h_u} = 1.102 \left[\frac{V_u}{V_c} \left(1 - \frac{h_w}{h_u} \right) \right]^{0.603}$$

Figure 4. Equation. Modified Umbrell et al. equation for maximum equilibrium scour.[2]

Lyn reanalyzed the data collected by Arneson and Abt and Umbrell et al.[1–3] Lyn identified concerns related to spurious correlation in the regression for the Arneson and Abt equation and the low quality of Umbrell et al.'s dataset. He proposed the empirical power law formulation in figure 5.

$$\frac{y_s}{h_u} = min \left[0.105 \left(\frac{V_b}{V_c} \right)^{2.95}, 0.5 \right]$$

Figure 5. Equation. Lyn equation for maximum equilibrium scour.[3]

The equation in figure 1 is currently recommended for use in the FHWA guidance document for bridge scour at highway bridges.[4] Concerns raised by Lyn suggest that an improved model for pressure scour is needed.[3]

CHAPTER 3. MODEL FORMULATION FOR PRESSURE FLOW SCOUR

Experiments show that scour is generally greatest near the downstream end of the bridge deck. This observation is commonly attributed to the vertical contraction (concentration) of flow in the bridge opening. The flow separates from the leading edge of the bridge deck, creating a flow separation zone (ineffective flow area) and forcing the flow in the bridge opening to contract and accelerate. High velocities in the bridge opening initiate sediment movement and scour.

INTRODUCTION

Pressure flow (vertical contraction) scour may be analyzed by estimating the effective depth in the bridge opening for critical velocity to occur at equilibrium scour. The effective depth (contraction depth) is determined based on the opening that allows stream flow to pass through at the location of maximum scour. Figure 6 illustrates pressure scour and defines key parameters including y_s and effective depth at the point of maximum scour, y_2.

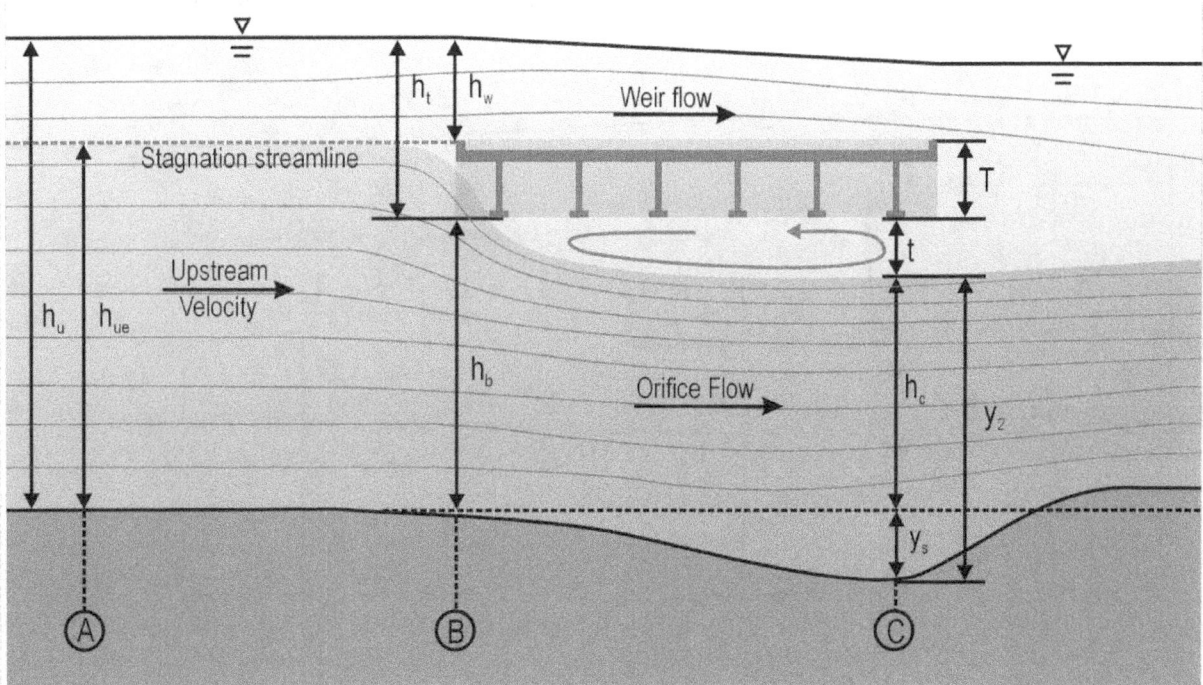

Figure 6. Illustration. Parameter definitions at maximum scour.

If h_u is greater than h_b, then there is a vertical contraction of flow that may result in scour. Figure 6 illustrates a fully submerged bridge deck where flow overtops the bridge. There is a stagnation streamline that represents the division of the approach flow between that which passes under the bridge and that which overtops the bridge. The effective approach flow depth, h_{ue}, represents the portion of the approach flow that is directed under the bridge. For partially submerged bridge decks, the bridge deck blocks the approach flow, but there is no weir flow over the deck, and all of the approach flow passes under the bridge.

5

At section C in figure 6, separation zone thickness, t, is indicated where there is effectively no flow conveyance. The flow field is effectively conveyed through the opening represented by the sum of flow contraction depth at the point of maximum scour, h_c, and y_s.

Figure 7 illustrates velocity distributions at three section locations. At the approach section A, the velocity distribution is unaffected by the bridge. For the fully submerged case shown in the figure, a portion of the approach flow will overtop the bridge, and a portion will flow beneath the bridge. The flow at the bridge opening, which is shown in section B, is non-uniform, which indicates a velocity peak (highest velocity) just below the flow separation zone. The velocity profile at the point of maximum scour, which is shown in section C, is also non-uniform, which indicates the possibility of minor backflow in the separation zone. For analytical purposes, actual velocity distributions are simplified to average velocities, as shown in figure 8.

As shown in figure 7, shear stress applied by the flowing water, τ_o, is less than the critical shear stress, τ_c, in the approach, as is required for clear water approach conditions. However, if there is sufficient contraction, this relationship will reverse at section B, initiating scour. At section C, a scour hole will form to increase conveyance until the applied shear is less than or equal to τ_c.

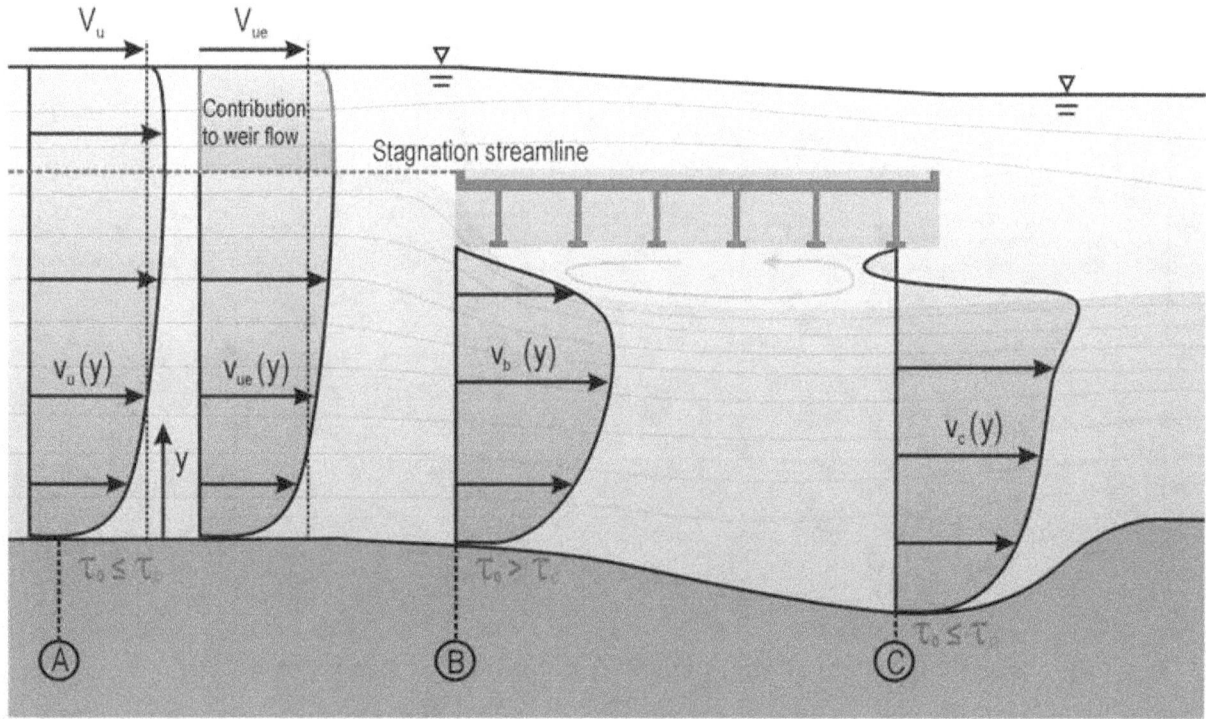

Figure 7. Illustration. Velocity distributions.

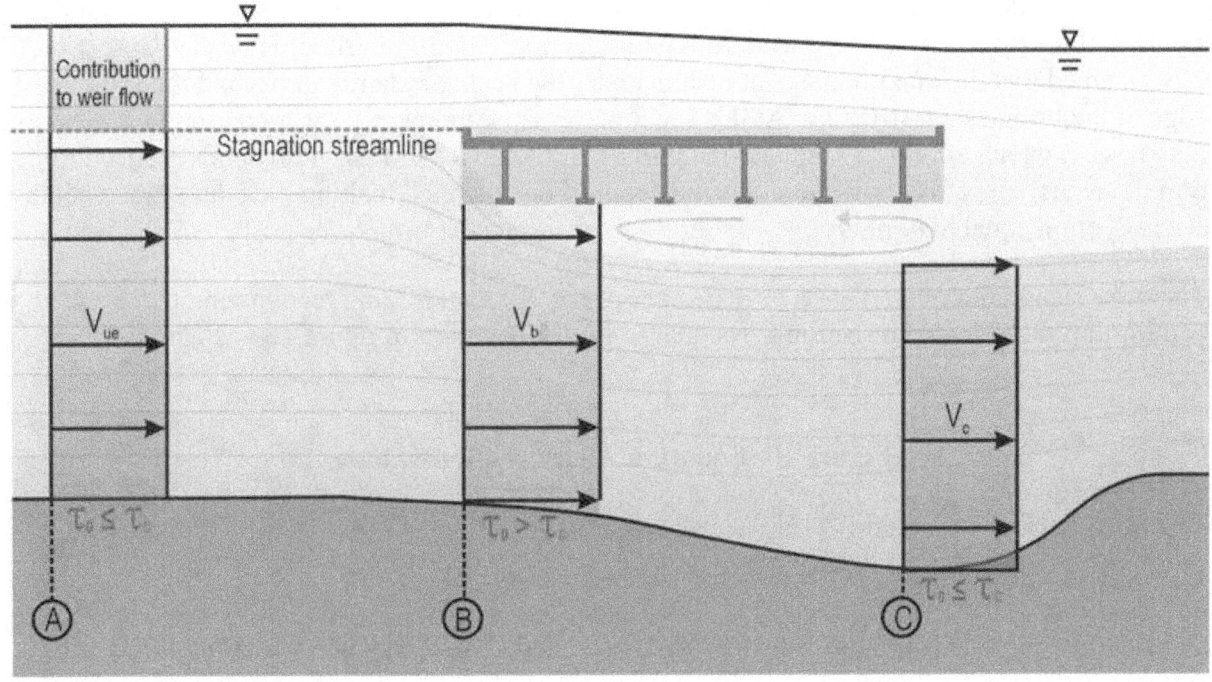

Figure 8. Illustration. Average velocities.

CONSERVATION OF MASS

Contraction of the flow field is created by a partially or fully submerged bridge superstructure because of the physical blockage of the superstructure resulting in contraction of the flow. The contraction creates a separation zone that starts from the lower corner of the leading edge of the bridge superstructure and increases to a maximum *t*. Determination of the flow separation thickness is considered critical to understanding the pressure flow scour mechanism.

Analysis of pressure flow scour focuses on the orifice flow under the bridge as described in figure 6. By the principle of conservation of mass, the discharge rate for the orifice flow must remain the same for sections A, B, and C. The discharge for unit width of channel is given by the equation in figure 9.

$$q = V h$$

Figure 9. Equation. Unit discharge.

Where:

q = Unit discharge, ft^2/s.
V = Average flow velocity, ft/s.
h = Flow depth, ft.

The zone of separation near the lower flange of the girders is filled with vortices, and the average velocity in this zone is nearly zero. In figure 6, section A is located sufficiently upstream so that it is free of the influence from the bridge superstructure. h_{ue} is measured from the stagnation point elevation to the unscoured stream bed. Section B is located at the leading edge of the bridge superstructure and is considered to be the beginning of the separation layer. The depth of

7

the flow in section B is h_b. Although some pressure scour may occur here, it is not sufficiently significant to add to the conveyance through this section. Maximum scour is located at section C. It is assumed that the maximum scour occurs under the bridge rather than beyond the trailing edge of the bridge superstructure. At this section, the separation zone has increased to a thickness, t, which effectively reduces the portion of the flow field providing conveyance to h_c. Simultaneously, y_s effectively increases the depth of the channel, making the conveyance depth at this section equal to h_c plus y_s.

Therefore, using the conservation of mass at sections A, B, and C and recognizing that at equilibrium the velocity at section C is equal to V_c, the equation in figure 10 is created.

$$V_{ue}h_{ue} = V_b h_b = V_c(h_c + y_s)$$

Figure 10. Equation. Conservation of mass.

Where:

V_{ue} = Effective approach velocity directed under the bridge, ft.

Considering only the continuity between sections A and C and recognizing that $h_c = h_b - t$, the equation in figure 11 is created.

$$V_{ue}h_{ue} = V_c(h_b - t + y_s)$$

Figure 11. Equation. Conservation of mass between sections A and C.

When the velocity in the contracted section C is less than the critical velocity of the bed material, scour will stop (or not begin). Conversely, when the velocity in the contracted section is greater than critical velocity, scour will occur, and the depth of scour will increase until the conveyance increases to the point where the velocity is reduced to the critical velocity. This occurs at the equilibrium scour depth. Laursen's relation for critical velocity is shown in figure 12.[5]

$$V_c = K_U(h)^{1/6}D_{50}^{1/3}$$

Figure 12. Equation. Laursen's critical velocity.[5]

Where:

K_U = Constant equal to 11.17 ft^2/s.

Substituting the active flow depth at section C ($h_b - t + y_s$) in Laursen's equation, Laursen's equation for V_c in figure 12 yields the equation in figure 13.[5]

$$V_{ue}h_{ue} = K_U(h_b - t + y_s)^{1/6} D_{50}^{1/3}(h_b - t + y_s)$$

Figure 13. Equation. Conservation of mass at equilibrium scour.

Solving this for equilibrium scour results in the equation in figure 14. The first term on the right-hand side of the equation is y_2, as shown in figure 6.

$$h_b + y_s = \left(\frac{V_{ue} h_{ue}}{K_U D_{50}^{1/3}} \right)^{6/7} + t$$

Figure 14. Equation. Equilibrium scour.

This relationship reveals that the equilibrium depth of scour is a function of the unit discharge through the bridge opening, $V_{ue} h_{ue}$, D_{50}, and t. By inspection, scour will only occur when the equation in figure 15 is satisfied.

$$\left(\frac{V_{ue} h_{ue}}{K_U D_{50}^{1/3}} \right)^{6/7} + t \geq h_b$$

Figure 15. Equation. Threshold for scour.

The values of h_b and D_{50} are determined from the site geometry and sampling of bed materials, respectively. Determination of the effective approach flow conditions and t require further consideration.

EFFECTIVE FLOW

The depth of the effective approach flow, h_{ue}, requires a determination of the location of the stagnation streamline. The location of the streamline will vary, but it is expected to be between the lowest point on the bridge superstructure and near or slightly above the top of the bridge railing. The location of the stagnation streamline is attributed to a number of factors, which may include the shape of the deck, pitch angle of the deck (super elevation), inundation depth, and weir flow depth. The presence of railing and size of openings in the railing may also affect the stagnation point.

Consider the case where the approach flow elevation does not overtop the bridge but any further increase in depth will result in overtopping. At this level of incipient overtopping, all discharge goes under the bridge deck, and the stagnation line is effectively at the top of the bridge superstructure. When the upstream depth exceeds the top of the superstructure, a small weir flow occurs. With negligible weir flow, it is reasonable to assume that for this boundary event, the stagnation level is very close to the top of the bridge superstructure. However, as the approach depth further increases, it is expected that the stagnation level will gradually move to some intermediate level on the side of the bridge superstructure as it splits the flow to weir flow above the bridge and orifice flow below the bridge. In most cases, the blocked flow directed below the bridge will be larger than the blocked flow over the bridge, allowing the hypothesis that the stagnation level will be located in the upper half of the superstructure.

Given the broad range of possible situations, there is no known theoretical means for determining the general separation streamline location. This is explored further with the experimental tests, but for the initial formulation of the scour model, it is assumed that the separation streamline is located at the top of the bridge superstructure as shown in figure 6.

With this assumption, the approach flow depth directed under the bridge is defined by the equation in figure 16.

$$h_{ue} = h_u = h_b + h_t$$

Figure 16. Equation. Estimate of effective approach flow depth.

Where:

h_t = Flow depth above the bottom of the bridge superstructure, ft.

For partially submerged flow, h_t is less than or equal to the bridge superstructure thickness, T, and for partially submerged flow, h_t is less than or equal to T. For fully submerged flow, h_{ue} represents the portion of the approach flow that will be directed under the bridge deck. This has been defined as the distance between the stagnation streamline and the unscoured streambed. As previously stated, the stagnation streamline is conservatively assumed to be at the top of the superstructure. Therefore, the effective approach depth for fully inundated flow is estimated by the equation in figure 17.

$$h_{ue} = h_b + T$$

Figure 17. Equation. Estimate of h_{ue} for fully inundated flow.

The effective approach velocity for V_{ue} must also be estimated. For a partially submerged deck, all flow is directed under the bridge, and V_{ue} is equal to V_u. For the fully submerged situation, V_{ue} is estimated using the power law velocity profile as shown in the equation in figure 18.

$$V_{ue} = V_u \left(\frac{h_{ue}}{h_u}\right)^n$$

Figure 18. Equation. Power law estimate of effective approach velocity.

Where:

n = Exponent used for the power law velocity distribution.

For fully developed turbulent flows, n may be taken as one-seventh. In the field, the assumption of fully developed turbulent flow is reasonable, but the same is not always true for laboratory (flume) data. A comparison between a fully developed profile and a uniform profile is shown in figure 19. Because of this difference in velocity profile, the power law exponent should be taken as zero for experimental data.

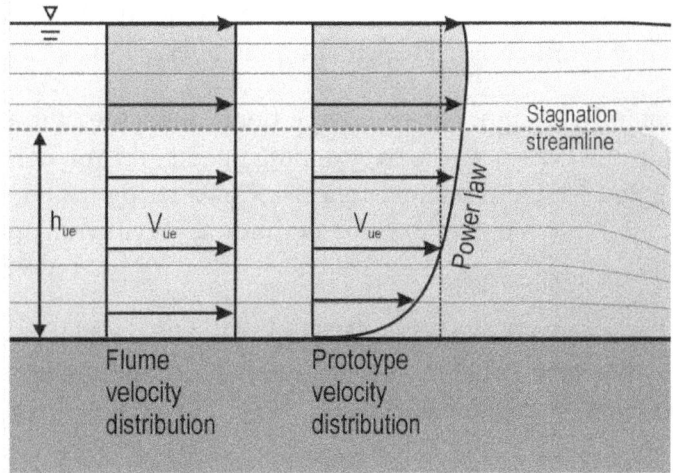

Figure 19. Illustration. Comparison of fully developed and uniform velocity profiles.

SEPARATION ZONE THICKNESS

The remaining parameter in the equation in figure 14 is t. Its value is not readily available from experimental data; therefore, it is necessary to develop an analytical model that ties t to other measured parameters. t is potentially affected by several factors, including the superstructure geometry, vertical flow contraction, bridge opening size, and approach flow conditions.

The relative importance of these parameters in determining t varies depending on the relationship between the bridge superstructure and the two flow boundaries—the channel bottom and the water surface. Therefore, it is useful to consider four conditions characterized by the relative position of bridge superstructure, the channel bottom, and the water surface. Assuming a fully submerged bridge superstructure, the conditions are as follows:

1. In the first condition, significant overtopping occurs, and the superstructure is relatively close to the stream bed. Conveyance over the bridge is large compared to the constriction caused by the superstructure. The bottom of the superstructure is sufficiently close to the channel bottom, and the distance between the two may be relevant to t. This condition is referred to as "surface-far-field."

2. In the second condition, moderate to insignificant overtopping occurs, and the superstructure is well above the stream bed. Conveyance under the bridge is large compared to the constriction caused by the superstructure. However, the proximity of the water surface to the separation zone is important in t. This condition is referred to as "bed-far-field."

3. In the third condition, significant overtopping occurs, and the superstructure is well above the stream bed. Conveyance both above and below the bridge is significant compared to the constriction caused by the superstructure. In this case, neither the water surface nor channel bed influences t. This condition is referred to as "far-field."

4. In the fourth condition, moderate to insignificant overtopping occurs, and the superstructure is relatively close to the stream bed. The constriction caused by the superstructure is

11

significant compared to the conveyance both over and under the bridge. Both the water surface and channel bed influence t. This condition is referred to as "shallow water."

Because bridges are generally designed to pass large flood flows (i.e., the 100-year event) without overtopping, for most cases where overtopping occurs, it is moderate to insignificant. Therefore, conditions 1 and 4 are of practical concern. Considering the partially submerged superstructure, conditions 1 and 3 do not exist by definition, leaving only conditions 2 and 4 for consideration.

For condition 2, t is a function of the characteristics of water, the extent of overtopping, and the constriction imposed by the superstructure. Conceptually, the equation in figure 20 describes a function that relates relevant parameters to t.

$$f(t, g, \mu, \rho, h_t, q_B) = 0$$

Figure 20. Equation. Parameters relevant for _t_.

Where:

μ = Viscosity of water, lb-s/ft^2.
ρ = Density of water, slug/ft^3.
q_B = Unit discharge blocked by the bridge superstructure, ft^2/s.

The unit discharge blocked by the bridge superstructure is based on V_u and the physical blockage of the superstructure. For submerged flow, q_B is defined in figure 21. For partially submerged flow, it is defined in figure 22.

$$q_B = V_u T$$

Figure 21. Equation. Unit discharge blocked for fully submerged flow.

$$q_B = V_u h_t$$

Figure 22. Equation. Unit discharge blocked for partially submerged flow.

For submerged flow, h_t is greater than T, and for partially submerged flow, h_t is less than or equal to T. With all potential variations of bridge geometry, general geometrical characterization of the bridge upstream face is a complex task. It is assumed that the tested bridge shapes are fairly representative for the largest population of typical bridges.

For submerged flow, the parameters described in figure 20 were configured into dimensionless ratios as shown by the equation in figure 23.

$$\frac{t}{h_t} = f\left(\frac{q_B}{\sqrt{g}h_t^{3/2}}, \frac{\rho q_B h_t}{T\mu}\right)$$

Figure 23. Equation. Dimensionless parameter ratios for submerged flow.

For partially submerged flow, the parameters described in figure 20 were configured into dimensionless ratios as shown by the equation in figure 24.

$$\frac{t}{h_t} = f\left(\frac{q_B}{\sqrt{g}h_t^{3/2}}, \frac{\rho q_B}{\mu}\right)$$

Figure 24. Equation. Dimensionless parameter ratios for partially submerged flow.

Substituting the definition of blocked discharge from the equation in figure 21, the equation in figure 23 becomes the equation in figure 25.

$$\frac{t}{h_t} = f\left(\frac{V_u}{\sqrt{gh_t}}\frac{T}{h_t}, \frac{\rho V_u h_t}{\mu}\right)$$

Figure 25. Equation. Revised dimensionless parameter ratios for fully submerged flow.

Similarly, substituting the definition of blocked discharge for partially submerged flow from figure 22, the equation in figure 24 becomes the equation in figure 26.

$$\frac{t}{h_t} = f\left(\frac{V_u}{\sqrt{gh_t}}, \frac{\rho V_u h_t}{\mu}\right)$$

Figure 26. Equation. Revised dimensionless parameter ratios for partially submerged flow.

The only difference between the fully and partially submerged conditions is presence of the ratio of T to h_t. For the fully submerged case, h_t is the sum of T and h_w, leading to the relationship shown in figure 27.

$$\frac{T}{h_t} = 1 - \frac{h_w}{h_t}$$

Figure 27. Equation. Submerged depth relationship.

Recognizing that in the partially submerged case, h_w is zero, the ratio in figure 27 will always be equal to 1 under such conditions. Therefore, both the fully and partially submerged cases are represented by figure 28.

$$\frac{t}{h_t} = f\left(\frac{V_u}{\sqrt{gh_t}}\left(1 - \frac{h_w}{h_t}\right), \frac{\rho V_u h_t}{\mu}\right)$$

Figure 28. Equation. Unified dimensionless parameter ratios.

Considering the shallow water condition (condition 4), additional parameters have an effect on t, including the clearance between the bottom of the bridge superstructure and h_b. Unlike condition 2, the bridge superstructure is sufficiently close to the stream bed so that the relative constriction influences t. By observation from experiments and CFD simulations, this proximity makes the separation zone thinner. Such an effect is more pronounced when the deck is very close to the bed.

For the shallow water condition, two cases must be satisfied. First, t should approach zero when the bottom of the superstructure approaches the water surface elevation. That is, t should approach zero when h_t approaches zero. Second, t should approach zero when the bottom of the superstructure approaches the stream bed. That is, t should approach zero when h_b approaches zero. While the latter limit is informative, such a situation is not realistic because a bridge would not be constructed so close to the stream bed and the scour depth would not necessarily approach zero in the limit. Nevertheless, these two limits can be expressed as dimensionless ratios by dividing by h_u and supplementing the parameters in figure 28 with these additional terms to develop the proposed equation for t in figure 29.

$$\frac{t}{h_t} = K \left(\frac{V_u}{\sqrt{gh_t}} \left(1 - \frac{h_w}{h_t} \right) \right)^m \left(\frac{\rho V_u h_t}{\mu} \right)^n \left(\frac{h_b}{h_u} \right)^p \left(\frac{h_t}{h_u} \right)^q$$

Figure 29. Equation. Equation form for t.

Where:

K = Variable related to bridge superstructure geometry, dimensionless.

The variation of K with different geometry is beyond the scope of this study. A few typical deck-and-stringer sections were selected and used for both experimental and CFD studies.

The two components of the first term of the equation in figure 29 can be separated as shown in figure 30.

$$\frac{t}{h_t} = K \left(\frac{V_u}{\sqrt{gh_t}} \right)^m \left(\frac{\rho V_u h_t}{\mu} \right)^n \left(\frac{h_b}{h_u} \right)^p \left(\frac{h_t}{h_u} \right)^q \left(1 - \frac{h_w}{h_t} \right)^m$$

Figure 30. Equation. Revised equation form for t.

CHAPTER 4. EXPERIMENTAL STUDY

New pressure flow scour experiments were conducted at FHWA's TFHRC J. Sterling Jones Hydraulics Laboratory to further develop and test the hypotheses proposed in this study. In addition, data from Arneson and Abt as well as Umbrell et al. were retrieved and used in this study.[1,2] In addition, a series of CFD simulations were conducted to expand the scope of the experimental data.

TFHRC EXPERIMENTS, PIV, AND CFD ANALYSES

Physical experiments were conducted in a flume under controlled flow conditions for two uniform bed materials and two bridge deck models. The facilities, instrumentation, experimental setup, and procedure are described in this section. The experimental flume was 70 ft long, 6 ft wide, and 1.8 ft deep with clear sides and a stainless steel bottom with a slope of 0.0007 percent (see figure 31). A test section that consisted of a narrowed channel that was 10 ft long and 2.07 ft wide and had a 1.3-ft sediment recess was installed in the middle of the flume. A model bridge was installed in the narrowed section above the sediment recess. A honeycomb flow straightener and a trumpet-shaped inlet were carefully designed to smoothly guide the flow into the test channel. Water was supplied by a circulation system with a sump of 7,400 ft^3 and a pump with capacity of 10.6 ft^3/s with the flow depth controlled by a tailgate. The discharge was controlled by a LabView^{TM} program and checked by an electromagnetic flow meter.

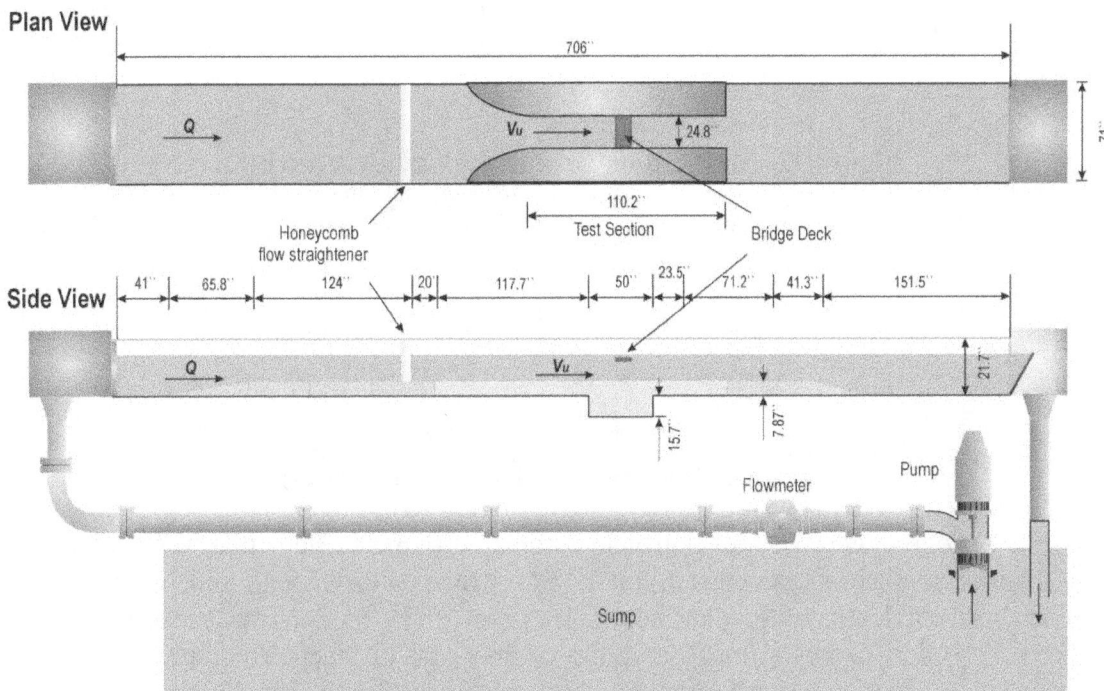

Figure 31. Illustration. Experimental flume.

To test the effect of sediment size on scour morphology, two uniform sands (gradation coefficient $\sigma < 1.5$) were used in the experiments: (1) $D_{50} = 0.0512$ inches and $\sigma = 1.45$ and (2) $D_{50} = 0.0858$ inches and $\sigma = 1.35$.

The effect of bridge girder configuration was examined using a three-girder deck and a six-girder deck as shown in figure 32. The bridge model was located at eight different elevations. Both decks had adjustable rails that could pass overflow on the deck surface, as shown in figure 33, permitting the deck to have eight different inundation levels.

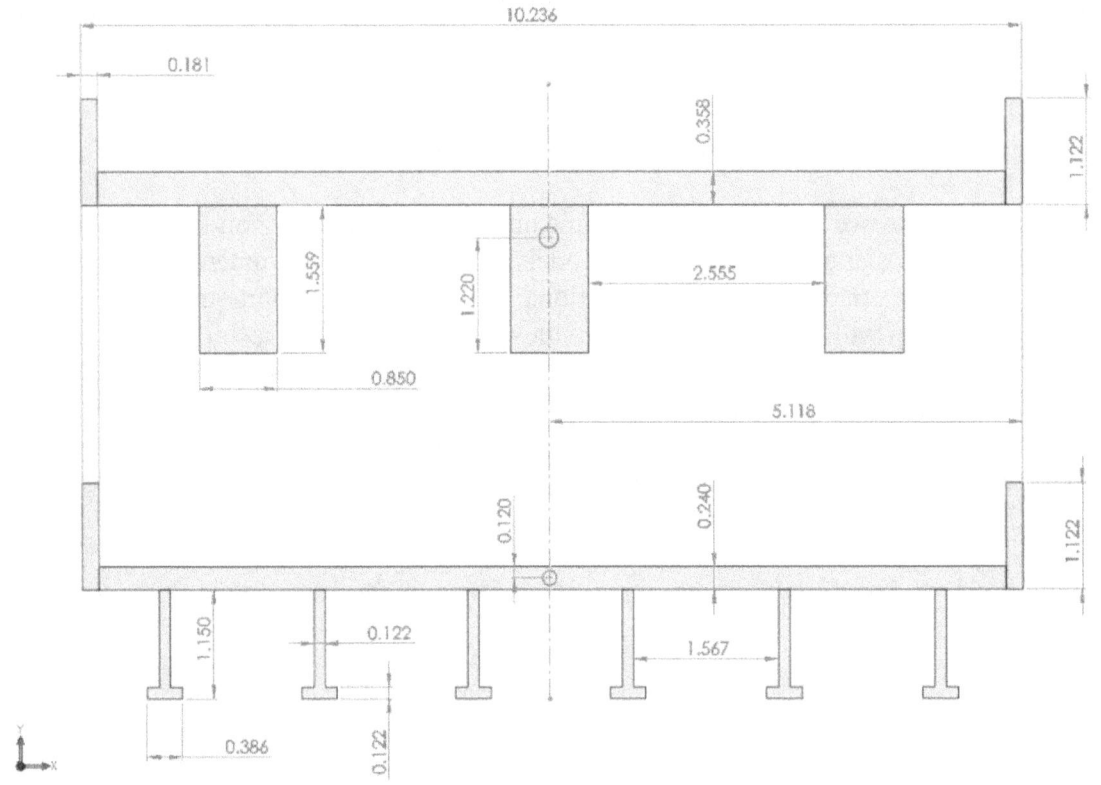

Figure 32. Illustration. Bridge deck models (inches).

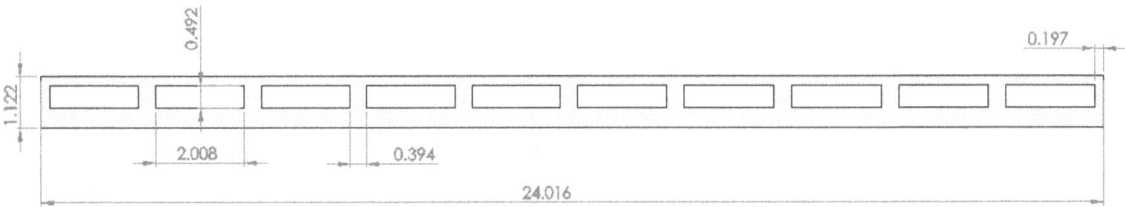

Figure 33. Illustration. Bridge rail (inches).

A LabView[TM] program was used to control an automated flume carriage that was equipped with a micro-acoustic doppler velocimeter (MicroADV) for records of velocities and a laser distance sensor for records of depths of flow and scour. The MicroADV by SonTek® measures three-dimensional (3D) flow in a cylindrical sampling volume of 0.177 inches in diameter and 0.220 inches in height with a small sampling volume located about 0.2 inches from the probe.[6] The range of velocity measurements is from about 0.0033 to 8.2 ft/s. For these experiments, velocity measurements were taken in a horizontal plane located at a cross section 0.72 ft upstream of the bridge model. The LabView[TM] program was set to read the MicroADV probe and the laser distance sensor for 60 s at a scan rate of 25 Hz. According to *Acoustic Doppler Velocimeter Technical Documentation, Version 4.0*, the MicroADV has an accuracy of

±1 percent of measured velocity, and the laser distance sensor has an accuracy of ±0.00787 inches.[6]

Two discharges were applied. They were determined by the critical velocity and the flow cross section in the test channel with a constant flow depth of 0.82 ft. The critical velocity was preliminarily calculated by Neill's equation and was adjusted downward by approximately 10 percent for the flow velocity to insure clear water scour.[7] The upstream velocity for the smaller of the two bed materials was approximately 1.3 ft/s, and the corresponding discharge was 2.28 ft³/s. With a Reynolds number of 57,000 and Froude number of 0.17, this approach flow was subcritical turbulent flow. The upstream velocity for the larger bed material was approximately 1.74 ft/s, and the corresponding discharge was 2.95 ft³/s. With a Reynolds number of 73,700 and Froude number of 0.22, this approach flow was also subcritical turbulent flow. The experimental conditions are summarized in table 3 in the appendix.

The experimental procedure is as follows:

1. Fill the sediment recess with sand and evenly distribute the sand on the bottom of the flume until the depth is 2 ft in the sediment recess and 0.66 ft in the test channel.

2. Install a bridge deck at a designated elevation and position it perpendicular to the direction of flow.

3. Pump water gradually from the sump to the flume to the experimental discharge that is checked with the electromagnetic flow meter.

4. Run each test for 36 to 48 h and monitor scour processes by grades in a clear side wall. An equilibrium state is attained when scour changes at a reference point are less than 0.0394 inches continuously for 3 h.

5. Gradually empty water from the flume and scan the 3D scour morphology using the laser distance sensor with a grid size of 1.97 by 1.97 inches.

PIV was employed to assist in the evaluation of t in a separate set of experiments in a smaller flume. A PIV system generally includes a laser emitter, charge coupled device camera, and reflective particles that follow the flow. The flow is first seeded with PIV particles (median diameter of 3.15 mil) upstream of the test section. A laser beam is spread into a light sheet by a cylindrical lens and illuminates the particles that travel across the width of the light sheet. Two images are taken by the camera with a short time delay.[8] The recorded images are then divided into small interrogation windows. The displacement of the particles in each interrogation window and subsequently the velocity can be obtained using cross correlation techniques. The bridge deck used in the PIV experiments is 0.583 ft wide, 0.909 ft long, and 0.148 ft high. The height of the girders is 0.066 ft.

CFD simulation was used to extend the laboratory data collected as part of this study. The CFD models of the flume were calibrated to the physical model runs, and additional scenarios could be generated. STAR-CCM+ CFD software was installed at the Transportation Research and Analysis Computing Center at the Argonne National Laboratory and was used to conduct simulations to estimate the boundary layer thickness and analyze the stagnation point of deck blockage for partially and fully submerged bridge deck flow conditions.[9] Flow fields adjacent to a representative bridge deck for unscoured and scoured bed were simulated. The six-girder bridge deck used for CFD simulations had the same dimensions as the model bridge deck used for the flume tests.

ARNESON AND ABT'S EXPERIMENTS

Arneson and Abt's experiments were conducted at Colorado State University in a 300-ft-long by 9-ft-wide flume with a bed slope adjustable up to 2 percent.[1] Maximum flow circulation capacity was 100 ft^3/s. Data were collected with various flow conditions, four separate bed materials, and six model bridge deck elevations. Arneson and Abt conducted a total of 72 tests. The experimental setup, measurements, and data description are detailed in *Transportation Research Record 1647*.[1] A summary of the test conditions is presented in table 4 of the appendix.

For the purposes of this study, some data points collected by Arneson and Abt were not considered. For the current study, data were required to meet the following criteria:

- Only clear water scour was considered.

- The bridge model low chord elevation resulted in perceptible vertical contraction.

- Measured scour was at least 0.394 inches.

Clear water scour occurred when V_u was less than V_c, as determined by Laursen's critical velocity equation shown in figure 12 using D_{50} as the characteristic sediment size.[5] Test runs with maximum scour less than 0.394 inches were discarded because the results would not be considered meaningful by introducing spurious correlations.

UMBRELL'S EXPERIMENTS

The experiments of Umbrell et al. were conducted at TFHRC in the same flume shown in figure 31.[2] Data were collected with various flow conditions, three separate bed materials, and six model bridge deck elevations. Umbrell et al. conducted a total of 81 tests. The experimental setup, measurements, and data description are detailed in "Clear-Water Contraction Scour Under Bridges in Pressure Flow."[2] A summary of the test conditions is presented in table 5 of the appendix.

As with the Arneson and Abt data, some data points collected by Umbrell et al. were not considered. Data were required to satisfy the following criteria:

- Only clear water scour was considered.

- The bridge model low chord elevation resulted in perceptible vertical contraction.

- Measured scour was at least 0.394 inches.

Clear water scour occurred when V_u was less than V_c, as determined by Laursen's critical velocity as described previously.[5] Test runs with maximum scour less than 0.394 inches were discarded.

CHAPTER 5. ANALYTICAL MODEL DEVELOPMENT

The analytical model described in figure 30 was further developed and tested using the data from the physical flume tests as well as through the use of flow visualization using PIV and CFD modeling.

SEPARATION ZONE THICKNESS

CFD and PIV analyses were used to assist in the characterization of the separation zone. Figure 34 and figure 35 provide examples of the CFD and PIV simulations, respectively. Scour rate was initially highest near the bridge opening, but as the scour hole developed, the deepest scour and largest t moved close to the downstream end of the bridge opening, as shown in both figures.

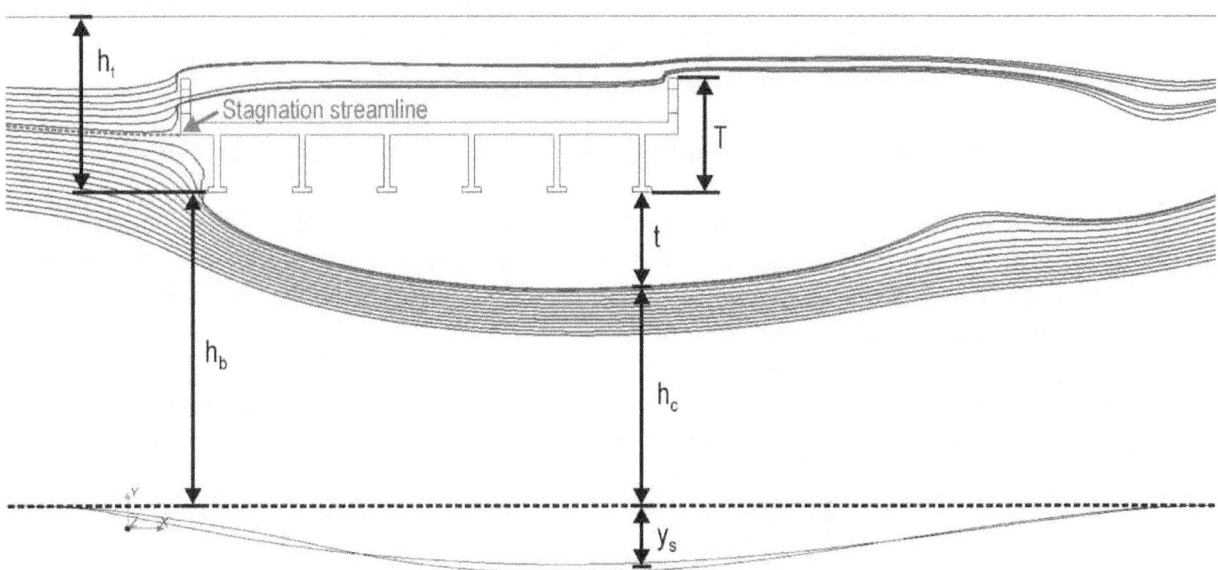

Figure 34. Illustration. Streamlines and separation zone from CFD simulation.

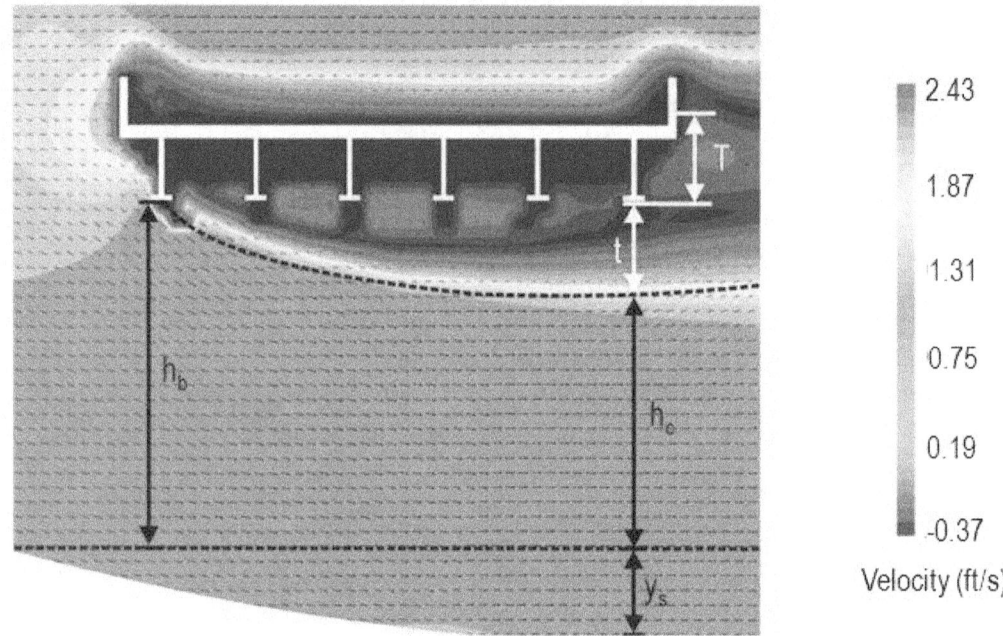

Figure 35. Illustration. Streamlines and separation zone from PIV data.

CFD simulations revealed that t was consistent with the model length scale. That is, the ratio of the first and second exponents in figure 30, m/n, was 2. Ignoring the slight geometric difference of deck models used in the experimental program, this was confirmed in the curve fitting for the values of K and the exponents shown in figure 36.

It should be noted that within the range of 41 to 68 °F, the viscosity of water decreased by approximately one-third. A temperature of 68 °F was assumed for the laboratory data used in this study.

$$\frac{t}{h_t} = 0.8\left(\frac{V_u}{\sqrt{gh_t}}\right)^{-0.3}\left(\frac{\rho V_u h_t}{\mu}\right)^{-0.15}\left(\frac{h_b}{h_u}\right)^{1.2}\left(\frac{h_t}{h_u}\right)^{-0.8}\left(1-\frac{h_w}{h_t}\right)^{-0.3}$$

Figure 36. Equation. Best fit t equation.

Using figure 36 in conjunction with figure 14, the scour depth can be calculated. This computation, using the laboratory data and normalized by upstream approach depth, is summarized in figure 37. Since the exponents and coefficient for estimating t were fit to these data, the fit is relatively good.

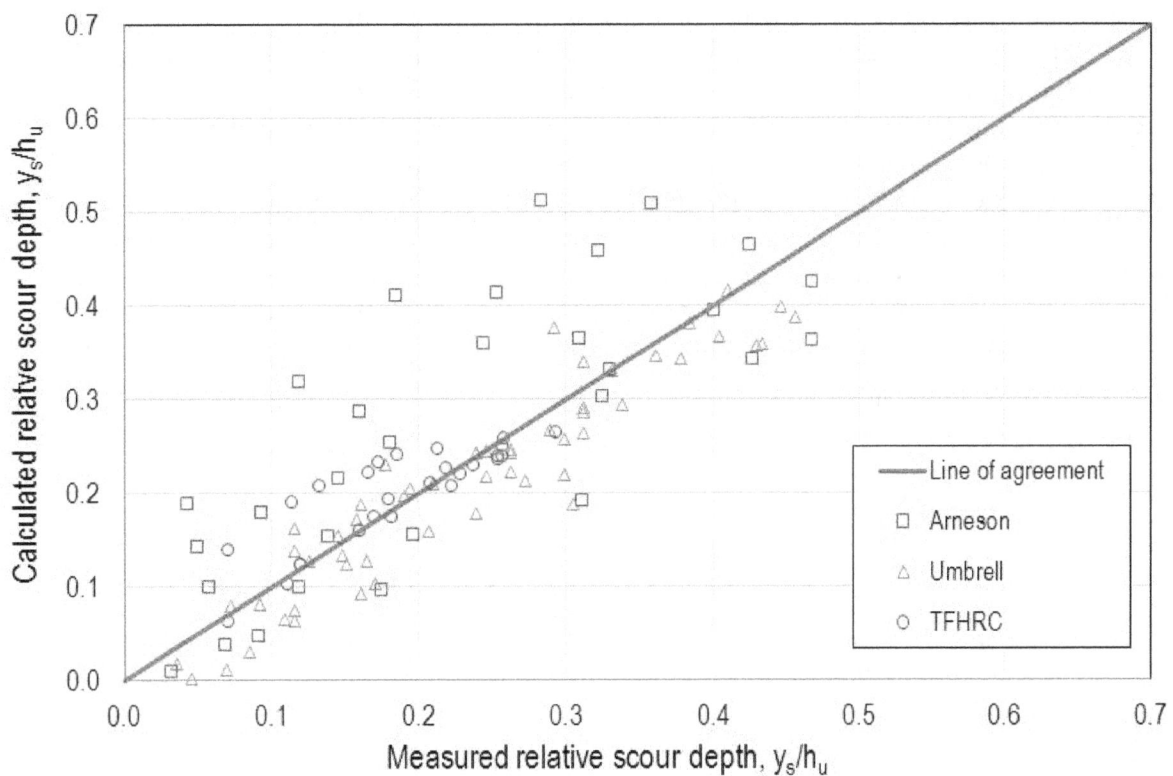

Figure 37. Graph. Scour comparison with best fit equation for *t*.

Because of the complexity of the equation for *t*, further assessment is desirable. Given that the ratio of the exponents on the first and second terms in figure 36 is 2, the effect of h_t in the first term cancels out the effect of that variable in the second term. Therefore, any length parameter may be substituted without changing the result of the equation. Substituting h_u for h_t results in the equation in figure 38. The first term in this equation is the approach Froude number, and the second term is the approach Reynolds number.

$$\frac{t}{h_t} = 0.8 \left(\frac{V_u}{\sqrt{gh_u}} \right)^{-0.3} \left(\frac{\rho V_u h_u}{\mu} \right)^{-0.15} \left(\frac{h_b}{h_u} \right)^{1.2} \left(\frac{h_t}{h_u} \right)^{-0.8} \left(1 - \frac{h_w}{h_t} \right)^{-0.3}$$

Figure 38. Equation. Best fit *t* with approach depth.

The estimation of *t* in figure 38 can be simplified by multiplying both sides of the equation by h_t/h_b, yielding the equation in figure 39.

$$\frac{t}{h_b} = 0.8 \left(\frac{V_u}{\sqrt{gh_u}} \right)^{-0.3} \left(\frac{\rho V_u h_u}{\mu} \right)^{-0.15} \left(\frac{h_b h_t}{h_u^2} \right)^{0.2} \left(1 - \frac{h_w}{h_t} \right)^{-0.3}$$

Figure 39. Equation. Best fit *t* with superstructure height.

However, CFD tests revealed behavior of *t* that was inconsistent with the formulation provided in figure 39. First, these tests indicated that *t* increased with approach depth for partially submerged conditions but became nearly constant after the bridge was fully submerged.

23

In contrast, the last term of the equation suggests that t would continue to increase. Figure 40 illustrates CFD simulations that show the increase in t with increasing approach depth. In contrast, figure 41 shows that as the partial submergence approached the top of the superstructure and became fully submerged, t was nearly constant. (Note that both figures show simulations with an approach velocity of 4.9 ft/s.) The cases are for a fixed bed, and, consequently, the thickest point of the separation zone is close to the upstream opening of the flooded bridge. The fixed bed imposes some constriction to the separation zone, making it narrow down and reattach sooner than when there is a scour hole. It was found that the initial expansion of the separation zone was not affected by this constriction. Therefore, an exponent closer to zero for this term is considered for the design equation, changing it from -0.3 to -0.1.

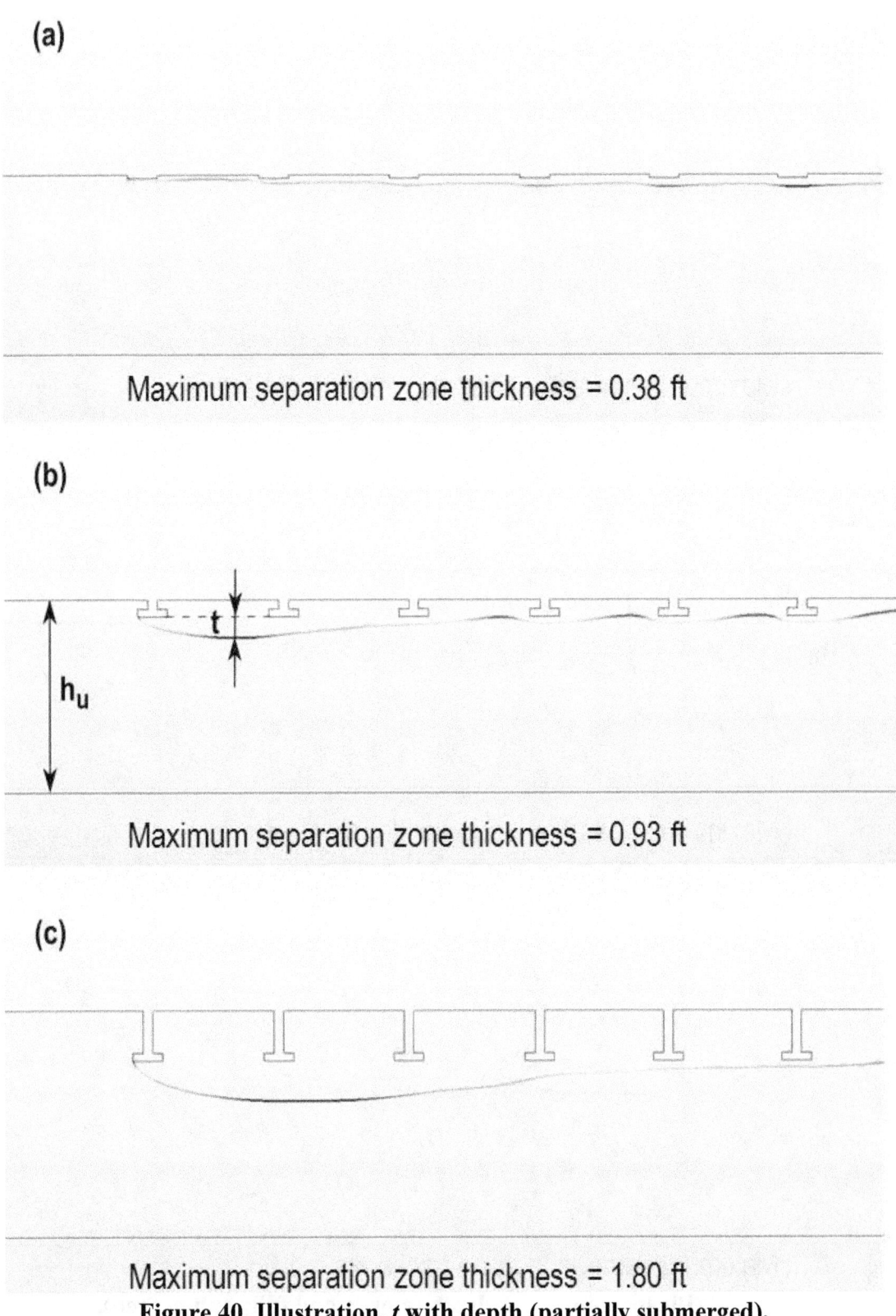

(a)

Maximum separation zone thickness = 0.38 ft

(b)

Maximum separation zone thickness = 0.93 ft

(c)

Maximum separation zone thickness = 1.80 ft

Figure 40. Illustration. *t* **with depth (partially submerged).**

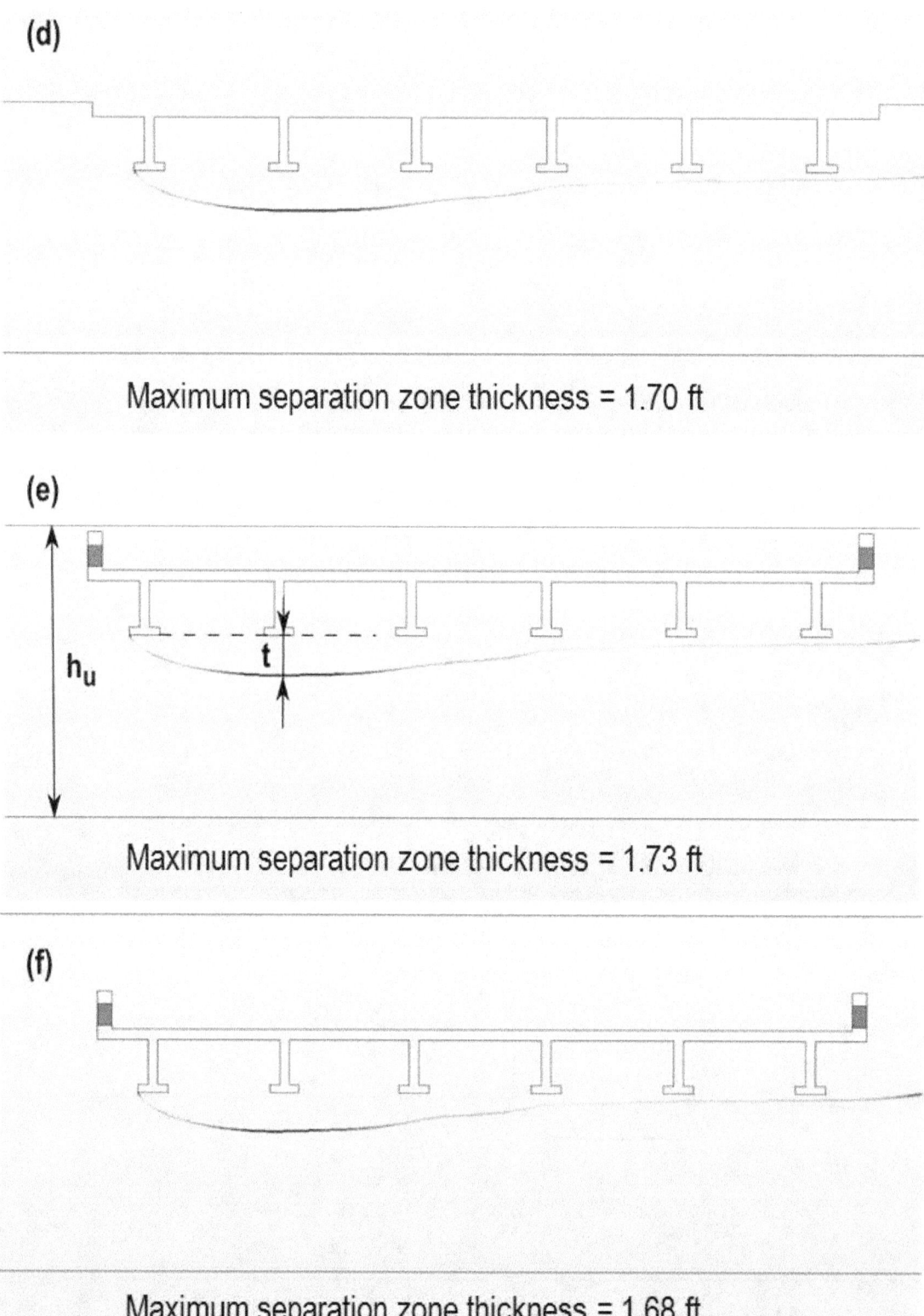

(d)

Maximum separation zone thickness = 1.70 ft

(e)

Maximum separation zone thickness = 1.73 ft

(f)

Maximum separation zone thickness = 1.68 ft

Figure 41. Illustration. *t* with depth (mostly and fully submerged).

The first two terms of the equation in figure 39 suggest a strong inverse relationship between approach velocity and t. However, CFD tests indicated that t did not decrease with approach velocity at the velocities anticipated at full scale. To systematically verify this, a CFD study was implemented on a full-scale bridge model with a clear span and with an upstream channel and bridge opening width of 40 ft, h_b of 7.87 ft, and a bridge superstructure height of 3 ft. The superstructure configuration was scaled up from the six-girder model in figure 32. A range of simulations were completed by varying both the approach depth and velocity, as shown in figure 42. Except at the lower end of the h_t/T ratio, the three velocity relationships are virtually identical. To address this, a reference velocity representative of the experimental datasets was used in the first two terms of the equation in figure 39 and then combined with the constant K, eliminating the strong velocity dependence. This revised value of K was approximately 0.2. The result of these two changes is shown in figure 43.

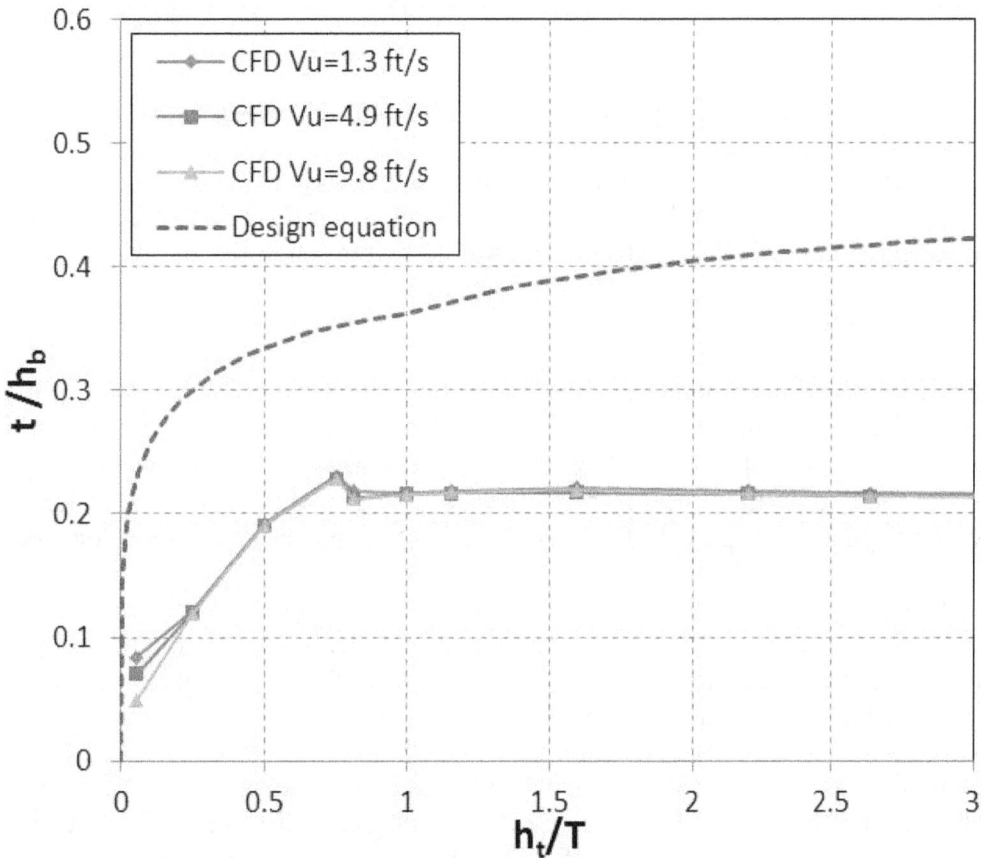

Figure 42. Graph. t in CFD tests.

$$\frac{t}{h_b} = 0.2\left(\frac{h_b h_t}{h_u^2}\right)^{0.2}\left(1 - \frac{h_w}{h_t}\right)^{-0.1}$$

Figure 43. Equation. Modified best fit equation for t.

Finally, the combined coefficient, K, was adjusted upward to provide a degree of conservatism to the model for design. The appropriate degree was defined as a reliability index of 2, where reliability index is defined by the equation in figure 44.

$$\beta = \frac{1 - \mu_Z}{\sigma_Z}$$

Figure 44. Equation. Reliability index.

Where:

β = Reliability index.
μ_Z = Mean of the variable Z.
σ_Z = Standard deviation of the variable Z.

The variable Z is defined as the ratio of the measured scour depth to the calculated scour depth. The resulting design equation is provided in figure 45.

$$\frac{t}{h_b} = 0.5 \left(\frac{h_b h_t}{h_u^2} \right)^{0.2} \left(1 - \frac{h_w}{h_t} \right)^{-0.1}$$

Figure 45. Equation. Proposed design equation for *t*.

This design equation for *t* is plotted in figure 42 for comparison with the CFD results for an unscoured full-scale bridge. The design follows the same general shape as the CFD observations while providing the desired conservatism as measured by the reliability index.

STAGNATION POINT

CFD modeling was used to evaluate the location of the stagnation point. Thus far, it has been assumed that the stagnation point for fully submerged flow is at the top of the bridge superstructure. CFD simulations of fully inundated decks showed that the stagnation point was generally near the mid-point of the superstructure. This is shown in figure 46 for the case of a rigid unscoured bed and in figure 47 for a noncohesive bed material after scour occurs.

The bridge superstructures modeled using CFD had large openings on the parapets simulating open bridge rails. Not all bridges will have such openings, or, if they do, the openings may be smaller. For bridges with smaller or no side openings above the pavement level, the stagnation point may be higher than indicated in the CFD analyses.

The bridge models used by Arneson and Abt as well as Umbrell et al. do not have parapet openings as the TFHRC models do, which may affect the stagnation streamline location.[1,2] Structure design variations create differences in results. With the data available, it is not yet possible to draw definitive conclusions about the stagnation point location. Therefore, the assumption that it is at the top of the bridge superstructure is maintained.

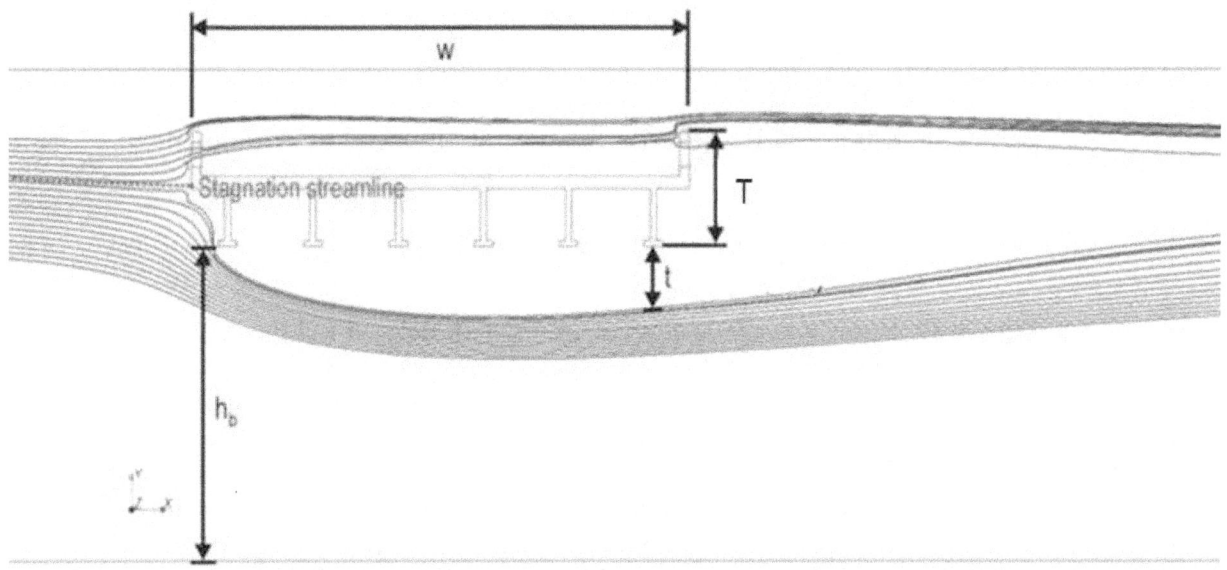

Figure 46. Illustration. Streamlines for unscoured bed.

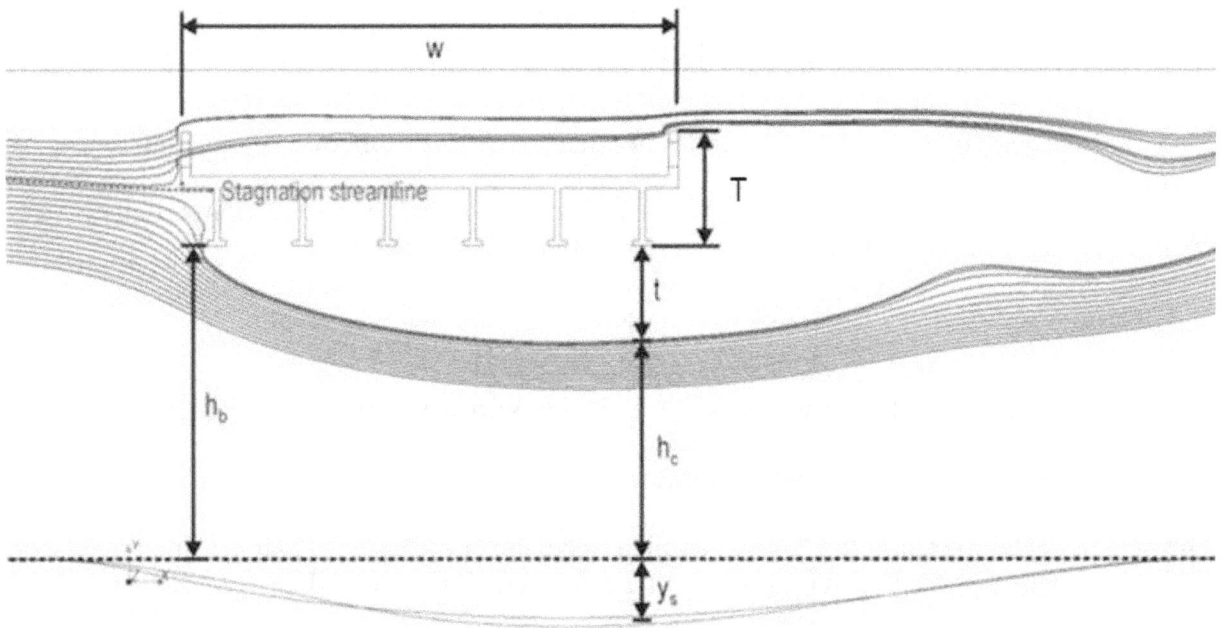

Figure 47. Illustration. Streamlines for scoured bed.

SCOUR DEPTH PREDICTION

Combining the general equation for scour in figure 14 with the equation for t in figure 43 yields the best fit scour equation in figure 48. Alternatively, using the t equation in figure 45 yields the proposed design equation in figure 49. Both equations are for fully or partially submerged bridge flow under clear water scour conditions.

$$y_s = \left(\frac{V_{ue} h_{ue}}{K_U D_{50}^{1/3}} \right)^{6/7} + \left[0.2 \left(\frac{h_b h_t}{h_u^2} \right)^{0.2} \left(1 - \frac{h_w}{h_t} \right)^{-0.1} - 1 \right] h_b$$

Figure 48. Equation. Best fit equation for submerged bridge flow.

$$y_s = \left(\frac{V_{ue} h_{ue}}{K_U D_{50}^{1/3}} \right)^{6/7} + \left[0.5 \left(\frac{h_b h_t}{h_u^2} \right)^{0.2} \left(1 - \frac{h_w}{h_t} \right)^{-0.1} - 1 \right] h_b$$

Figure 49. Equation. Design equation for submerged bridge flow.

A performance evaluation of the best fit and proposed design equations, including comparison with the Arneson and Abt method and with the method of Umbrell et al., is appropriate.[4,2] However, the Arneson and Abt and Umbrell et al. methods were based on analyses to provide best fits to the data being considered at the time. For this comparison, they were adjusted to consider a conservative envelope for design to directly compare with the proposed method. An optimization parameter, β, was added to the Arneson and Abt equation (from figure 1) to evaluate and compare its performance as shown in figure 50.[1]

$$\frac{y_s}{h_u} = -5.08 + 1.27 \left(\frac{h_u}{h_b} \right) + 4.44 \left(\frac{h_b}{h_u} \right) + 0.19 \left(\frac{V_b}{V_c} \right) + \beta$$

Figure 50. Equation. Arneson and Abt equation with optimization parameter.[1]

Similarly, an optimization parameter, α, was added to the Umbrell et al. equation in figure 4 to evaluate and compare its performance as shown in figure 51.[2]

$$\frac{y_s + h_b}{h_u} = 1.102 \left[\frac{V_u}{V_{uc}} \left(1 - \frac{h_w}{h_u} \right) \right]^{0.603} + \alpha$$

Figure 51. Equation. Umbrell et al. equation with optimization parameter.[2]

Performance Comparison of Best Fit Equations

The best fit equation from figure 48 was applied to the experimental data and is summarized in figure 52. These data result in a root mean square error (RMSE) of 0.070.

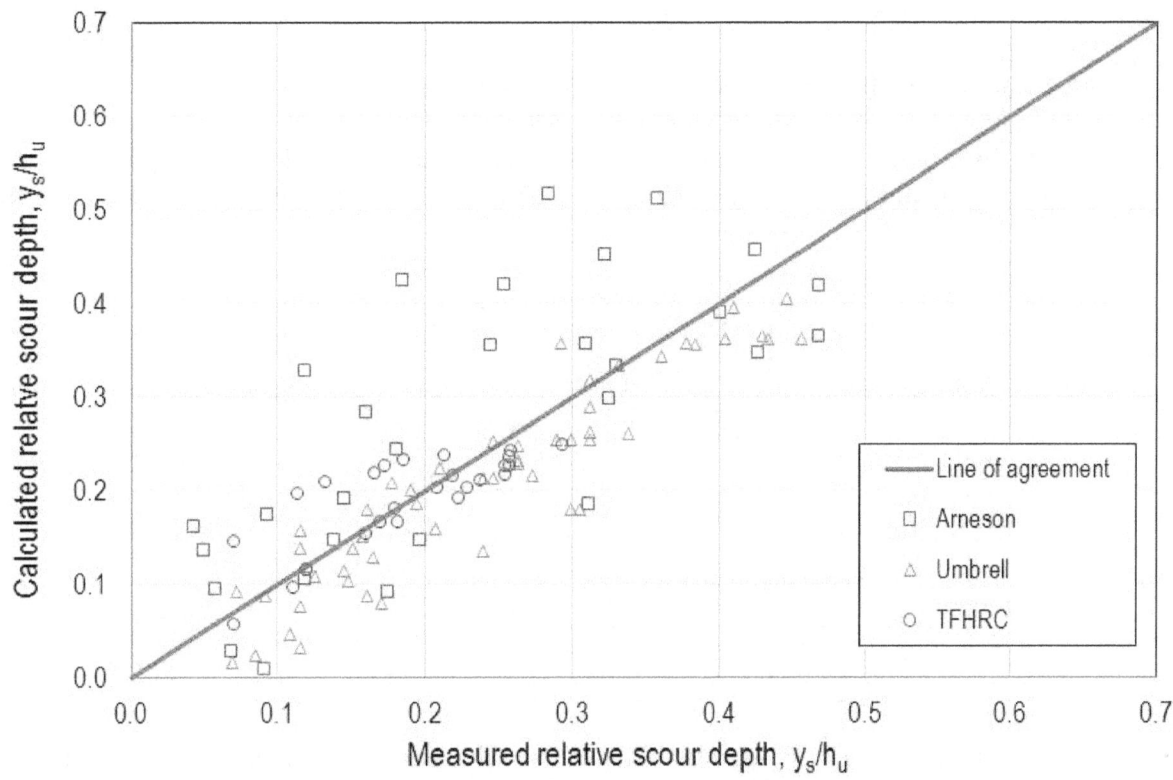

Figure 52. Graph. Scour comparison with best fit equation.

A comparison of the best fit model with the Arneson and Abt as well as Umbrell et al. equations is summarized in table 1.[1,2] The table includes the previously defined reliability index. The best fit and Arneson and Abt models exhibited a low reliability index, which is to be expected for best fit equations. The Umbrell et al. equation showed a higher reliability index, suggesting that some degree of conservatism was included in the model.

The RMSE of the 109 data points is also provided in table 1, along with the probability that the calculated estimate over predicts the observed scour. Both of these metrics are based on the dimensionless ratio of y_s/h_u.

Table 1. Summary of best fit metrics.

Comparative Metric	Best Fit Model (figure 48)	Arneson and Abt Model ($\beta = 0.0$)	Umbrell et al. Model ($\alpha = 0.0$)
Reliability index	-0.1	-0.8	0.6
RMSE of all observations	0.070	0.255	0.091
Probability of over prediction (percent)	34	32	80

Performance Comparison of Design Equations

The proposed design model was compared to the Arneson and Abt and Umbrell et al. models.[1,2] To provide a common basis for comparison, the parameters β and α were optimized so that the reliability index was close to 2, which was also performed when developing the proposed design model. The comparative metrics are summarized in table 2. With a reliability index of 1.9 for all models, the probability of overprediction ranged from 97 to 99 percent.

Table 2. Comparison of performance for design models.

Comparative Metric	Best Fit (figure 49)	Arneson and Abt Model ($\beta = 0.44$)	Umbrell et al. Model ($\alpha = 0.06$)
Reliability index	1.9	1.9	1.9
RMSE of all observations	0.153	0.457	0.136
RMSE of all over predictions	0.155	0.451	0.136
Probability of over prediction (percent)	97	97	99

The RMSE measures complement the reliability index. Given the equivalent reliability index and probability of over prediction, a lower RMSE is desirable. Based on RMSE, the Umbrell et al. model performs slightly better than the proposed model, with α equal to 0.06. The Arneson and Abt model performs significantly worse than either of the other two formulations.[1]

Performance is graphically represented in figure 53 for the proposed model, figure 54 for the Arneson and Abt model, and figure 55 for the Umbrell et al. model.[1,2] The graphs for the proposed model and the Umbrell et al. model are similar, as indicated by the RMSE values.

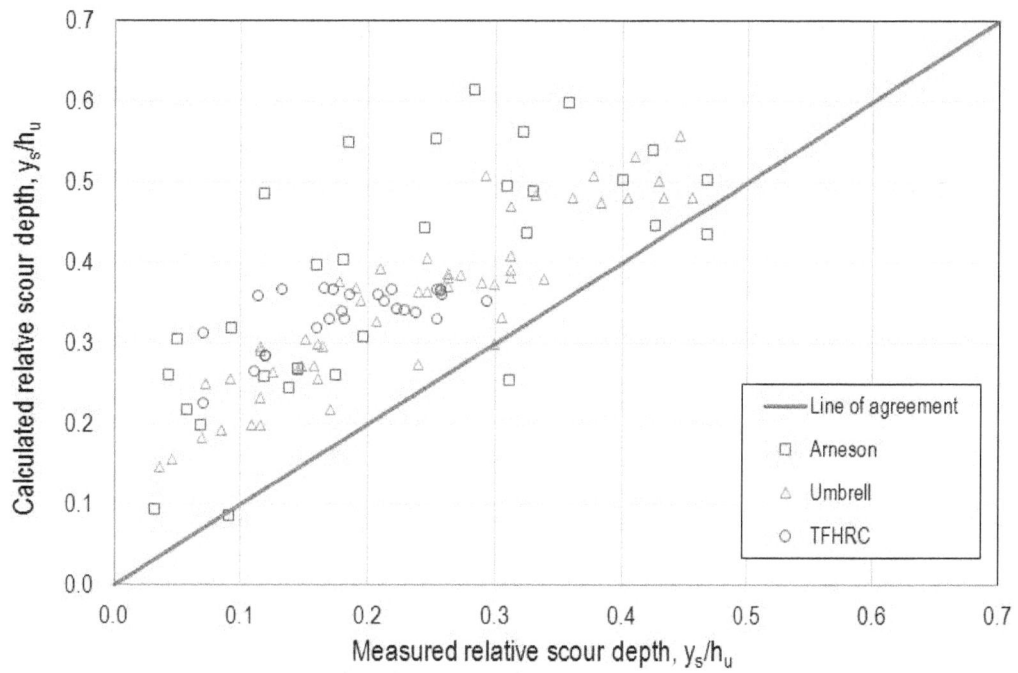

Figure 53. Graph. Performance of proposed model.

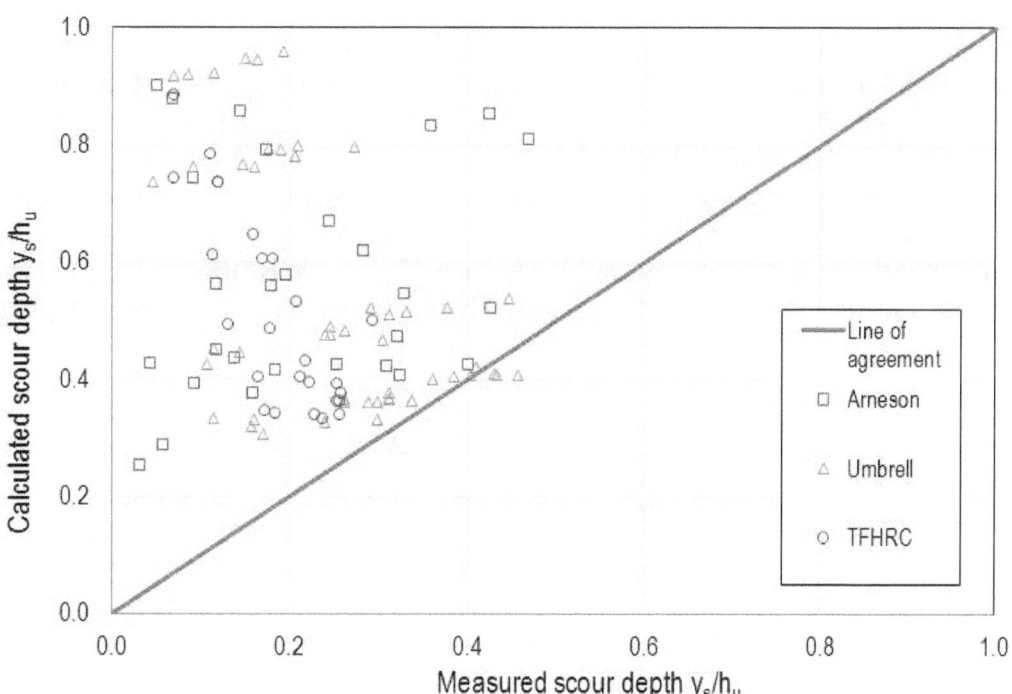

Figure 54. Graph. Performance of Arneson and Abt model ($\beta = 0.44$).

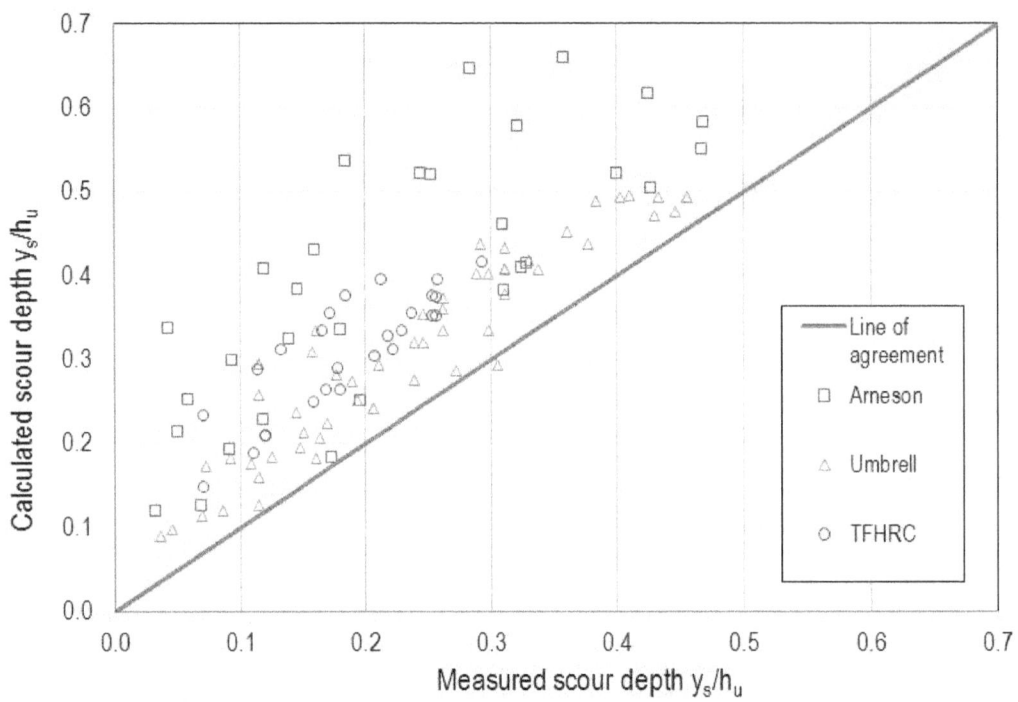

Figure 55. Graph. Performance of Umbrell et al. model ($\alpha = 0.06$).

Alternative methods of optimizing each model might be explored. However, the robustness of a model for application to data not used for its development depends on sound formulation based on physical principals. The Arneson and Abt model is primarily regression-based and does not perform as well as the Umbrell et al. model and proposed model on the combined laboratory data sets. The Umbrell et al. model is conceptually based on an analog to horizontal contraction scour. The proposed model is soundly based on physical principles as previously described. While the Umbrell et al. model performs slightly better when considering the RMSE for design, the physical basis of the proposed model should provide for robust application to field applications.

CHAPTER 6. CONCLUSIONS

This study was performed to develop a new model for estimating pressure flow (vertical contraction) scour under partially or fully submerged bridge superstructures. Based on this model, a proposed design equation for bridge foundation design for pressure scour is proposed. As part of this study, a set of laboratory experiments supplemented with PIV tests and CFD modeling were conducted. These data, combined with those from previous research efforts, were used to develop and validate the proposed scour model in clear water conditions with non-cohesive bed materials.

The laboratory experiments showed that under clear water conditions, the maximum scour occurred at the downstream end of the bridge deck. The maximum scour depth increased with deck inundation but decreased with increasing sediment size. It was also observed that the maximum scour depth was independent of the number of deck girders.

Scaled flume experiments, PIV tests, and CFD modeling confirmed that the maximum scour depth can be described by the effective depth (contracted depth) in the bridge opening at equilibrium scour. In order to evaluate the effective depth, it is necessary to estimate the boundary layer separation thickness from the leading edge of the bridge deck. When the bridge superstructure was lowered into flowing water, the leading edge developed a flow separation area. When the lower cord of the girder approached the bed or when the deck approached the water surface, the flow separation zone below the bridge deck was influenced by the bottom boundary and water surface, respectively.

Aided by a parametric CFD study, a design equation calculating t was developed and incorporated into an overall pressure (vertical contraction) scour model for design. Important features of this model include the following:

- Unified format for partial and full inundation.

- Reasonable fit to the three sets of laboratory data.

- Dimensionally scalable to the field.

- Conservative design values as defined by the reliability index.

APPENDIX. DATABASES

This appendix contains tabular summaries of the TFHRC laboratory data (table 3), Arneson and Abt laboratory data (table 4), and Umbrell et al. laboratory data (table 5).

Table 3. Summary of TFHRC pressure scour tests.

Test	D_{50} (inches)	Gradation σ	T (ft)	h_b (ft)	h_u (ft)	V_u (ft/s)	y_s (ft)
1	0.045	1.45	0.223	0.722	0.82	1.345	0.057
2	0.045	1.45	0.223	0.689	0.82	1.345	0.091
3	0.045	1.45	0.223	0.640	0.82	1.345	0.131
4	0.045	1.45	0.223	0.591	0.82	1.345	0.170
5	0.045	1.45	0.223	0.541	0.82	1.345	0.179
6	0.045	1.45	0.223	0.492	0.82	1.345	0.208
7	0.045	1.45	0.223	0.492	0.82	1.345	0.211
8	0.045	1.45	0.223	0.443	0.82	1.345	0.210
9	0.045	1.45	0.223	0.394	0.82	1.345	0.211
10	0.045	1.45	0.223	0.344	0.82	1.345	0.240
11	0.045	1.45	0.190	0.673	0.82	1.345	0.098
12	0.045	1.45	0.190	0.673	0.82	1.345	0.098
13	0.045	1.45	0.190	0.623	0.82	1.345	0.139
14	0.045	1.45	0.190	0.623	0.82	1.345	0.148
15	0.045	1.45	0.190	0.574	0.82	1.345	0.147
16	0.045	1.45	0.190	0.525	0.82	1.345	0.182
17	0.045	1.45	0.190	0.476	0.82	1.345	0.187
18	0.045	1.45	0.190	0.427	0.82	1.345	0.195
19	0.045	1.45	0.190	0.377	0.82	1.345	0.208
20	0.086	1.35	0.190	0.673	0.82	1.739	0.057
21	0.086	1.35	0.190	0.623	0.82	1.739	0.093
22	0.086	1.35	0.190	0.574	0.82	1.739	0.108
23	0.086	1.35	0.190	0.525	0.82	1.739	0.136
24	0.086	1.35	0.190	0.476	0.82	1.739	0.141
25	0.086	1.35	0.190	0.427	0.82	1.739	0.152
26	0.086	1.35	0.190	0.377	0.82	1.739	0.174

Table 4. Summary of Arneson and Abt pressure scour tests.[1]

Test	D_{50} (inches)	Gradation σ	T (ft)	h_b (ft)	h_u (ft)	V_u (ft/s)	y_s (ft)
6	0.035	2.24	0.830	0.540	1.590	0.490	0.145
8	0.035	2.24	0.830	1.250	1.420	1.310	0.096
9	0.035	2.24	0.830	1.080	1.440	1.300	0.282
10	0.035	2.24	0.830	0.910	1.470	1.370	0.477
11	0.035	2.24	0.830	0.740	1.490	1.350	0.596
12	0.035	2.24	0.830	0.570	1.550	1.330	0.725
14	0.035	2.24	0.830	1.273	1.510	1.371	0.262
15	0.035	2.24	0.830	1.110	1.570	1.610	0.517
28	0.130	2.23	0.830	0.810	1.510	1.325	0.087
29	0.130	2.23	0.830	0.640	1.500	1.234	0.207
30	0.130	2.23	0.830	0.470	1.560	1.235	0.484
33	0.130	2.23	0.830	0.990	1.430	1.728	0.169
34	0.130	2.23	0.830	0.820	1.510	2.309	0.278
35	0.130	2.23	0.830	0.650	1.510	1.828	0.644
36	0.130	2.23	0.830	0.480	1.560	1.846	0.729
40	0.024	1.58	0.830	0.810	1.410	0.588	0.045
41	0.024	1.58	0.830	0.640	1.480	0.720	0.063
42	0.024	1.58	0.830	0.470	1.390	0.674	0.202
44	0.024	1.58	0.830	1.240	1.410	1.322	0.070
45	0.024	1.58	0.830	1.070	1.470	1.270	0.265
46	0.024	1.58	0.830	0.900	1.470	1.287	0.454
47	0.024	1.58	0.830	0.730	1.520	1.265	0.489
48	0.024	1.58	0.830	0.560	1.540	1.226	0.653
64	0.059	3.00	0.830	0.900	1.400	1.380	0.130
65	0.059	3.00	0.830	0.730	1.460	1.359	0.233
66	0.059	3.00	0.830	0.560	1.440	1.353	0.352
69	0.059	3.00	0.830	1.060	1.480	1.898	0.175
70	0.059	3.00	0.830	0.890	1.520	1.853	0.384
71	0.059	3.00	0.830	0.720	1.690	1.951	0.479
72	0.059	3.00	0.830	0.650	1.740	1.978	0.623

Table 5. Summary of Umbrell et al. pressure scour tests.[2]

Test	D_{50} (inches)	Gradation σ	T (ft)	h_b (ft)	h_u (ft)	V_u (ft/s)	y_s (ft)
1	0.012	1.41	0.312	0.502	1.001	0.758	0.299
2	0.012	1.41	0.312	0.502	1.001	0.879	0.338
3	0.012	1.41	0.312	0.502	1.001	1.024	0.384
4	0.012	1.41	0.312	0.502	1.001	1.033	0.456
6	0.012	1.41	0.312	0.594	1.001	0.728	0.240
7	0.012	1.41	0.312	0.594	1.001	0.889	0.312
8	0.012	1.41	0.312	0.594	1.001	1.086	0.410
11	0.012	1.41	0.312	0.689	1.001	0.738	0.144
12	0.012	1.41	0.312	0.689	1.001	0.860	0.240
13	0.012	1.41	0.312	0.689	1.001	0.994	0.312
14	0.012	1.41	0.312	0.689	1.001	1.043	0.292
16	0.012	1.41	0.312	0.833	1.001	0.889	0.148
17	0.012	1.41	0.312	0.833	1.001	1.024	0.177
20	0.012	1.41	0.312	0.896	1.001	0.879	0.115
21	0.012	1.41	0.312	0.896	1.001	1.076	0.194
24	0.012	1.41	0.312	0.948	1.001	0.902	0.036
25	0.012	1.41	0.312	0.948	1.001	1.053	0.125
27	0.047	1.08	0.312	0.502	1.001	1.138	0.157
28	0.047	1.08	0.312	0.502	1.001	1.381	0.289
29	0.047	1.08	0.312	0.502	1.001	1.640	0.404
33	0.047	1.08	0.312	0.594	1.001	1.040	0.171
34	0.047	1.08	0.312	0.594	1.001	1.365	0.262
35	0.047	1.08	0.312	0.594	1.001	1.608	0.361
39	0.047	1.08	0.312	0.689	1.001	1.040	0.108
40	0.047	1.08	0.312	0.689	1.001	1.299	0.305
41	0.047	1.08	0.312	0.689	1.001	1.365	0.246
42	0.047	1.08	0.312	0.689	1.001	1.608	0.331
43	0.047	1.08	0.312	0.689	1.001	1.755	0.446
47	0.047	1.08	0.312	0.833	1.001	1.184	0.046
48	0.047	1.08	0.312	0.833	1.001	1.381	0.161
49	0.047	1.08	0.312	0.833	1.001	1.526	0.207
50	0.047	1.08	0.312	0.833	1.001	1.608	0.190
51	0.047	1.08	0.312	0.833	1.001	1.640	0.272
52	0.047	1.08	0.312	0.896	1.001	1.381	0.085
53	0.047	1.08	0.312	0.896	1.001	1.591	0.164
56	0.047	1.08	0.312	0.948	1.001	1.608	0.115
59	0.094	1.05	0.312	0.502	1.001	1.516	0.161
60	0.094	1.05	0.312	0.502	1.001	1.739	0.299
61	0.094	1.05	0.312	0.502	1.001	1.762	0.312
62	0.094	1.05	0.312	0.502	1.001	1.844	0.312
63	0.094	1.05	0.312	0.502	1.001	2.067	0.433
65	0.094	1.05	0.312	0.594	1.001	1.516	0.115

66	0.094	1.05	0.312	0.594	1.001	1.762	0.262
67	0.094	1.05	0.312	0.594	1.001	2.087	0.430
69	0.094	1.05	0.312	0.689	1.001	1.535	0.115
70	0.094	1.05	0.312	0.689	1.001	1.762	0.262
71	0.094	1.05	0.312	0.689	1.001	1.821	0.246
72	0.094	1.05	0.312	0.689	1.001	2.087	0.377
74	0.094	1.05	0.312	0.833	1.001	1.739	0.092
75	0.094	1.05	0.312	0.833	1.001	2.087	0.210
77	0.094	1.05	0.312	0.896	1.001	1.719	0.069
78	0.094	1.05	0.312	0.896	1.001	2.028	0.151
81	0.094	1.05	0.312	0.948	1.001	2.067	0.072

ACKNOWLEDGMENTS

This research was supported by the FHWA Hydraulics Research and Development Program with Contract No. DTFH61-11-D-00010. Roger Kilgore provided technical editing services.

REFERENCES

1. Arneson, L.A. and Abt, S.R. (1998). "Vertical Contraction Scour at Bridges with Water Flowing Under Pressure Conditions," *Transportation Research Record 1647*, 10–17.

2. Umbrell, E.R., Young, G.K., Stein, S.M., and Jones, J.S. (1998). "Clear-Water Contraction Scour Under Bridges in Pressure Flow," *Journal of Hydraulic Engineering, 124*(2), 236–240.

3. Lyn, D.A. (2008). "Pressure-Flow Scour: A Re-Examination of the HEC-18 Equation," *Journal of Hydraulic Engineering, 134*(7), 1015–1020.

4. Richardson, E.V. and Davis, S.R. (2001). *Evaluating Scour at Bridges*, 4th Edition, Hydraulic Engineering Circular No. 18, Report No. FHWA-NHI 01-001, Federal Highway Administration, Washington, DC.

5. Laursen, E.M. (1960). "Scour at Bridge Crossings," *Journal of the Hydraulics Division, 86*(HY 2), American Society of Civil Engineers, Reston, VA.

6. SonTek. (1997). *Acoustic Doppler Velocimeter Technical Documentation, Version 4.0*, SonTek, San Diego, CA.

7. Neill, C.R. (1973). *Guide to Bridge Hydraulics*, University of Toronto Press, Toronto, Ontario.

8. Shan, H. (2010). *Experimental Study on Incipient Motion of Non-Cohesive and Cohesive Sediments*, PhD dissertation, University of Nebraska-Lincoln, Omaha, NE.

9. Lottes, S.A., Bojanowski, C., Shen, J., Xie, Z., and Zhai, Y. (2012). *Computational Mechanics Research and Support for Aerodynamics and Hydraulics at TFHRC, Year 2 Quarter 1 Progress Report*, Argonne National Laboratory Report No. ANL/ESD/12-13, U.S. Department of Energy, Argonne, IL.

www.ingramcontent.com/pod-product-compliance
Lightning Source LLC
Chambersburg PA
CBHW080649180526
45168CB00008B/3349